LET THEM EAT CAKE

A NOVEL

by

TOM ANGUS

authorHOUSE™

1663 LIBERTY DRIVE, SUITE 200
BLOOMINGTON, INDIANA 47403
(800) 839-8640
WWW.AUTHORHOUSE.COM

© 2005 TOM ANGUS. All Rights Reserved.

No part of this book may be reproduced, stored in a retrieval system, or transmitted by any means without the written permission of the author.

First published by AuthorHouse 01/07/05

ISBN: 1-4208-1049-9 (e)
ISBN: 1-4208-1048-0 (sc)

Library of Congress Control Number: 2004098580

Printed in the United States of America
Bloomington, Indiana

This book is printed on acid-free paper.

This book is dedicated to the many fine scholars with whom the author has been privileged to work over the years.

Let Them Eat Cake
or
Qu'ils Mangent De La Brioche

Widely attributed to Marie Antoinette, Queen of France, who was alleged to have uttered these fateful words on being told that the people had no bread. She lived a life of luxury in a series of magnificent palaces and famously amused herself by dressing up as a shepherdess in the model village constructed for her in the park at Versailles. In reality she had no idea how ordinary people lived and had little contact with them. Then came the Revolution.

Whether or not Marie Antoinette ever said *Qu'ils mangent de la brioche*, the phrase has come to symbolize the indifference of a pampered elite to the real world beyond the palace gates.

Chapter 1: Klomp Goes Forth

He hated being called 'Klomp'. His name was Henri de Klompenmaker, but with typical linguistic laziness the London publishing world had decided that this was too much of a mouthful. Klomp grumpily hailed a taxi as he emerged from the terminal building at Bermuda International Airport. His mood belied the fact that he had been well entertained on the flight. The film, *'John Major, Man of Destiny'*, the life story of the grey man who had risen without trace to become Prime Minister of Great Britain, had been most inspiring. Like John Major, Henri de Klompenmaker's academic achievements were modest; like John Major, he had recently received a spectacular and unexpected promotion; like John Major, great things were now expected of him. In fact the whole flight had gone well until he had been held back to allow the First Class passengers to disembark. He couldn't believe his eyes. Marlene Pym in First Class! Of course he should have known that she would be flying to Bermuda for the Annual Retreat of the Newtonian Academy of Scholarly Publishing. He had been in London long enough to realise that no gathering of scholarly publishers was complete without Marlene Pym, marketing director of Albany Press, his own company's arch-rival. But First Class travel? That

1

was too much! Nobody at Standard International travelled first class, not even The Chairman – it was SI company policy. He looked the other way, but he could still hear her animated chatter as she moved in his direction. It got worse.

'*Mon dieu!*' she exclaimed in the affected French that was her linguistic signature. '*Mon dieu*, chaps, *c'est Klomp*!' Not only had she seen him, but she was obviously not alone. He turned, and found that Marlene and her companions had pretty well taken over the entire First Class compartment. Thus was Klomp's first flight in BA Club Class entirely ruined. Escorted by a fawning purser they walked past him. '*Bonjour, Klomp*', purred Marlene, with a mischievous flash of her dark eyes. 'Hello Klomp', said each of the six Englishmen who followed in her wake. He knew every one of them. They were all regulars at the quarterly luncheons of the London Publishers' Forum. The companies they worked for were so small! How could they justify First Class travel? Klomp's mood had deteriorated significantly by the time he and his fellow business-class passengers were allowed to disembark. When he reached immigration control he joined a long line waiting to be processed, but at least it did not include Pym and her gang; as First Class passengers they had gone through ahead of the rest. Nor, to his relief, was there any sign of them in baggage claim. Once he had collected his cases Klomp made his way to the taxi rank, having angrily rejected attempts to persuade him into a 'hotel shuttle'. He had heard that these were ruinously expensive and he would make do with a cab. His instinctive preference, of course, was for public transport, but he felt that this would not be in keeping with his status as a Company Director.

'Marlene Pym!' he brooded as the taxi wound its way along Bermuda's narrow lanes. He was still smarting from her attack on him at the last quarterly luncheon of the London Publishers' Forum. He had been invited to address them on his recent reorganization of SI's marketing department and was delighted to do so: he thought these dinosaurs had much to learn from him. Marlene Pym thought differently. Several of her friends had lost their jobs in his reorganization; Marlene Pym was fiercely loyal to her friends.

'Why have you fired all these experienced people?'

'Because we decided to outsource marketing. We don't regard it as a core competency.'

'Not core? Marketing? Then how do you sell your products?' Marlene Pym was not believing her ears.

'Marketing is about a lot more than selling products. SI is selling values and developing brand awareness.' This was, verbatim, what the management consultants who had recommended the reorganization had told him.

'That, I suppose, explains the 'Bravura Brands, Predictable Profits' full page advertisement in last week's *Financial Times,* which does not mention a single SI product.' Marlene's riposte prompted the first titters from the audience.

'Precisely.'

'And the 'Superior Solutions, Global Grasp' double page spread in the most recent issue of *Forbes Magazine*?

'Also an important strategic message'

'Is alliteration your only marketing tool?'

Once 'alliteration' had been explained to him, Klomp delivered his crushing reply. 'Of course not. As you're such an avid reader of the financial press, you should have seen '*SI*mply the Best' In the *Wall Street Journal* recently.'

'So you deploy hyperbole as well?'

He had no idea whether he did or not, but loud laughter from the Forum members prevented a further exchange of ideas.

Klomp didn't mind the questions so much; he had a textbook answer for every one. But he did mind the laughter they provoked. Henri de Klompenmaker did not like being made to look a fool. This is a lesson that Marlene Pym would have to learn. Perhaps he would have the opportunity to teach her in the coming days.

As Klomp's taxi drew up at the Devonshire Beach Resort, it was surrounded by porters eager to grab his cases. He fought them off, proud of his 'wheelie-bag' that allowed him to travel the globe entirely free of porter interference and minimised his outlay on tips (SI policy was NOT to reimburse tips). This had led to him being injured in a scrum at Karachi airport, but the company DID have global medical insurance cover. At the check–in desk he could hear the sound of merry banter from the direction of the bar; the same voices as on the plane. He decided to have dinner in his room that evening; he had seen enough of these people for one day. As he dragged his wheelie-bag through the moonlit gardens to his room he passed a porter laden with cases, many labelled Marlene Pym. The porter saw him surveying the enormous pile. 'They tip very well, sir', he said.

As was his habit when dining alone, Henri de Klompenmaker reviewed his triumphant career and raised a glass to the corporation that had nurtured it. He had indeed come a long way from that little Belgian town where he had been born and where he had grown up. And who would have predicted it? After high school, where the zenith of his achievement had been his captaincy of the ten-pin bowling team, he had progressed to a large university in the suburbs of Brussels. There he left no footprint other than his presidency of the stamp collectors' club, which he had founded and which, he understood, had not long outlived his departure. Despite great dedication to study, his grades had been unimpressive and his ambition to proceed to a Doctorate was thwarted. The summit of Klomp's academic achievement was, in consequence, a Master's degree. So be it. If the World of Learning did not want him he would dedicate himself to the World of Finance and he followed the family tradition by qualifying as an accountant at the *Ecole Europeenne de la Compatabilite*. Thus equipped, he sought profitable employment. To the embarrassment of his father, a lifelong Eurocrat, Klomp failed the examinations that would have gained him a place in the bureaucracy of the European Union, an endless supply of expense account lunches and, if he made it to the ripe old age of 55, a very generous pension. Alas, it was not to be, and a great bureaucratic talent was lost to Europe.

Rebuffed by the World of Government, Henri de Klompenmaker set his sights on the World of Business. He found a vehicle worthy of his ambition in the Standard International Corporation, the definitive multinational conglomerate and the nearest thing to the civil service in the business world. The fact that his grandfather had spent his entire career with SI, and that two of his uncles were still with it, smoothed his path and he was hired as an 'Executive, Global Accounts Payable' by its Elevator Division, based in Gent. Henri de Klompenmaker found that Standard International suited him perfectly and over the years he made steady progress through its financial backwaters.

SI had emerged from the murky world of the 19th Century Belgian Congo, where it had been granted extensive mining concessions by the notorious King Leopold. By the middle of the 20th Century, its mining activities had spread to every part of the world where labour was cheap and government corrupt. Inconvenient environmental lobbyists were not allowed to hamper SI's steady growth. Until the end of World War II, during which SI supplied both Allies and Axis with essential minerals, the company had never felt the need to look beyond mining for its profits. But what with the growing public concern about human rights, rising labour costs and the embarrassment of having several senior SI executives imprisoned for war crimes, the shareholders thought that some diversification might be prudent. The first step along this road was an easy one.

In 1946 a small, London-based publisher, Brogue & Co, founded in 1825 and one of the world's leading producers of geological maps, charts and atlases, had fallen on hard times. Strict post-war paper rationing meant that the high quality paper stock they required was impossible to obtain and despite healthy demand for Brogue charts, sales had plunged. Starved of cash and on the verge of bankruptcy, the company had no option but to put itself up for sale. SI, who had used their charts for almost a century, saw an opportunity. Initially Sir Algernon Brogue, Baronet, Chairman of Brogue and Co., was reluctant: he had hoped to sell the business to another publisher. But no other offer came close, and the family shareholders, now cash poor themselves, were insistent on the sale to SI. The purchase price made but a small dent in the SI gold reserves, housed in Switzerland, which had swelled hugely during the War.

SI's first act as owner was to decree that Brogue & Co. no longer publish its geological charts. Yes, their network of cartographers could still collect the data, and even create the charts, but from now on they would be available only to SI. Why feed the competition? Algernon Brogue, who had been allowed to stay on as Chairman, was outraged. He was only dissuaded from a very public resignation by being given a free hand to manage and develop Brogue & Co's small list of scholarly journals, which he had started in the 1930s. The fact that SI had the contacts to ensure that he, alone among journal publishers, would have access to limitless supplies of paper ensured that this part of the publishing business would flourish in the coming years. By the time Sir Algernon Brogue retired he had built up a publishing business that was admired on the trading floors of the City for its profits and in the groves of Academe for its quality.

Emboldened by the success of their first non-mining business, and with considerable Swiss gold reserves yet to be disposed of before awkward questions were asked, SI moved its corporate headquarters from Brussels to London and embarked on a series of acquisitions that transformed it into the organization whose tentacles now reach all parts of the industrial world. The Mining Division, once its *raison d'etre*, but now a growing political liability, was no more. It had been rebranded 'SI Natural Resources', which also encompassed forests, rivers and other wonders of nature that could be profitably exploited.

These days SI Publishing is probably best known for its stable of glossy consumer magazines, but the chief jewel in its crown remains Sir Algernon Brogue's collection of specialised scholarly journals, which now number over 500. No field of scholarly endeavour is too small or too obscure to have its very own journal. The twice-yearly issues of The *International Journal of Pre-Columbian Tribal Linguistics* are just as indispensable to the few, but passionate scholars in this field, as is each day's *Wall Street Journal* to the financiers of lower Manhattan. As a business scholarly journal publishing has many attractions. Chief among these is its profitability. Authors provide their articles for free, hoping that the visibility brought by publication will improve their prospects for promotion. Editors seek little or no payment, hoping that the visibility that comes with an editorship will enhance their status and increase their powers of patronage. But the real beneficiary of this system is the publisher, whose costs are, in contrast to his prices, remarkably low. Scholarly publishing may be rarefied, but it is highly and predictably profitable. Accountants love it.

As he sipped his solitary wine, Klomp's mind wandered back to the year of his Great Breakthrough. Nigel Archer MBA, the new SI Chairman, had been brought in to shake up the organization. After decades of steady growth, by the 1980s profits were stagnant and the share price was depressed. Most of SI's divisions were, as they say, 'underperforming'. Only the Publishing Division showed any growth, and the new Chairman decided that even more profits might be squeezed from it. To achieve this he decisively shifted the balance of power from the publishers - inconveniently creative types who were difficult to control - to the much more obedient accountants, whom he saw as his means of control. This not only undermined the publishing edifice so carefully constructed by Sir Algernon Brogue, but also transformed Henri de Klompenmaker, whose academic career had progressed no further than a Master's degree, into the Publishing Director responsible for SI's 500 scholarly journals.

Such a prize comes at a price, however, and the new Publishing Director's first task was to slash costs and reduce the size of his recently acquired empire. Far from being reluctant to wield the axe, Klomp was itching for an opportunity to use the tools he had discovered all those years ago at the *Ecole Europeenne de la Compatabilite*. Yes he, Henri de Klompenmaker, would show them the art of publishing by numbers. Not just numbers that told the mundane story of profit and loss, or of the pedantries of market share, but numbers, wonderful numbers that would reveal the inner workings of the academic world. Until now, such insights had been the exclusive preserve of the publishers, whose large population and extensive travel Klomp would cull. It had worked in other industries, even in healthcare. If numerical tools can substitute the medical judgement of doctors, why not of publishers? If a few, simple numbers can distinguish the good hospitals from the bad, why not journals? If accountants can take clinical decisions, why not publishing decisions?

Not all of SI's journal editors were convinced of the benefits of Klompean methods of management. Several, indeed, had resigned in protest. But in Professor Leslie Fyfe, Editor *ad interim* of *Transactions in Moleculetics*, one of SI's most profitable journals, Klomp had found a willing accomplice. The *ad interim* had lasted for several years, as Dr Fiona Hamilton, SI's inconveniently creative publisher responsible for *Transactions*, refused to make his position permanent and entertained hopes of removing Fyfe entirely. Eager to curry favour with her new boss, Fyfe had been a ready convert to Klomp's methods. Despite its financial success, *Transactions* was no longer a highly regarded publication – a fact that Fiona Hamilton attributed to the poor reputation of its Editor *ad interim*. Fyfe claimed that this was now changing and attributed the improvement to his adoption of Klompean methods. He supported this assertion with numerical data about some particularly highly cited articles from a laboratory somewhere in California. Les had also said something about important patents and sensational new weight loss treatments, but Henri de Klompenmaker did not concern himself with such details; he left the science to the Editor. As long as the numbers were right, Klomp was happy.

As for Fiona Hamilton, PhD; she was clearly a heretic, but had her uses. Even Klomp could not deny that she had launched a number of successful new journals, notably the triumphantly profitable co-publication with the Chinese, *Scripta Nanochimica Sinica.* The Chairman himself had stated at last year's corporate strategy meeting that China was the new frontier for SI, and Klomp was able to bathe in the reflected glory of Dr Hamilton's excellent relations with the orient. In any event, with the big cuts he had recently made in marketing, Klomp's staff reduction programme was on target, so she could stay for the moment. But next year further cuts would be required, and China would probably be out of fashion with the corporate strategists by then. Comforted by these thoughts, and aided by the bottle of wine that he had consumed, Henri de Klompenmaker slept soundly that night.

Chapter 2: The Newtonian Academy Retreats

Klomp rose at dawn the next morning, eager to face the challenges of the new day; the day on which he would dazzle the Fellows of the Newtonian Academy. Before going to breakfast he spent a couple of hours rehearsing his talk: '*A Quantitative Approach to Journal Management*'. He had been through it so many times that he now knew it virtually by heart. But he could leave nothing to chance: not with Marlene Pym in the audience. He read and re-read the text and checked and re-checked his slides. As he was using the journal *Transactions in Moleculetics* as his example, he had worked closely with Les Fyfe, its Editor *ad interim,* on the presentation. The results did appear to be spectacular. Since that momentous meeting at which Klomp had explained to Fyfe the realities of journal publishing, profits from *Transactions* were growing nicely, and it looked like they were now even attracting some decent papers, which would silence those, including Dr Fiona Hamilton, who said it was a low quality publication.

By 8.00 am Klomp was satisfied that his presentation was word perfect and went to breakfast. Now wide awake, he was able to take in his surroundings, which were indeed remarkably pleasant. The Devonshire Beach Resort, on 'Bermuda's sun-kissed south coast' was one of the oldest of the island's famous 'cottage colonies', self-contained resorts with every dining and sporting facility on site. In his tired and preoccupied state yesterday evening, he had not quite taken it all in. Now, as he walked to breakfast, he saw that the 'Cottage' containing his room was one of many scattered about lawns that swept down to the resort's private beach. Two storeys high, with four rooms in each, all had views of the ocean. His seemed to be furthest from the beach, but he assumed that the Committee would have the best rooms. He made a mental note to propose to The Chairman of SI that he, Henri de Klompenmaker, should relieve him of the chore of sitting on this Committee. He followed the signs to 'Newtonian Academy of Scholarly Publishing Breakfast' and his sense of satisfaction increased. Over to his left, he could see people frolicking in the surf, where a turquoise ocean broke upon a coral beach. To his right, two pre-breakfast doubles matches were in progress on the tennis courts, and the splashing sounds from the pool indicated some activity there. Apparently the elite of global academic publishing had found a variety of ways to prepare for the first day of the retreat, not all of them involving the study of papers in a bedroom.

To reach this sun-kissed Bermudian beach the Newtonian Academy of Scholarly Publishing had come a long way - strayed, according to its critics- from its original mission to 'Promote the Exchange of Speculations on Natural Philosophy Among Poor Scholars.' Founded in 1787 by two elderly Cambridge dons who had sat at the feet of the great Isaac Newton , its original endowment consisted of Appletree House, the home of one of its co-founders and the only non-Collegiate building on The Backs; and the collected papers of the other, together with the London home that housed them. Appletree House served as the Academy's headquarters and also provided accommodation for its 20 Fellows on their occasional visits. The collected papers provided sufficient material to fill the first four volumes of the Academy's *Proceedings*. The rent from the London house, once emptied of the collected papers, provided, along with the modest sales of the *Proceedings*, the Academy's only income. For the next 170 years the Newtonian Academy prospered quietly in its Cambridge backwater, its Bursar content that annual income invariably exceeded annual expenditure; its Editor content that the supply of Speculations on Natural Philosophy was more than sufficient to fill the quarterly issues of the *Proceedings*. All this changed in 1947, when the Academy ran out of funds. Rent from the London house had ceased to flow one night in 1941, when the house itself had ceased to exist, courtesy of the Luftwaffe. By 1945, publication of the *Proceedings* had also ceased, a victim of the paper rationing. By 1947 the Academy's once-ample cash reserves had been used up and it looked like Appletree House itself would have to be sold. Sir Algernon Brogue, hearing of its plight, approached the Academy with an offer that Brogue & Co., now owned by SI, should bail it out in return for the right to publish the *Proceedings*, but the fact that several senior SI executives in Belgium had recently been imprisoned for war crimes made the Fellowship reluctant to associate their Academy with that organization. It was at this point that one Captain Robert Maxwell, war hero, appeared on the scene. Eager to build up his new scholarly publishing business, he had heard of the Academy's difficulties, and one day swept into Appletree House armed with a chequebook. A bibulous luncheon was arranged with the Fellowship and a deal was done. Captain Maxwell would take over publication of the *Proceedings*, providing the Academy with a generous, guaranteed income in return. Better still, he offered to buy the ruin of their London house, provided they agreed to membership of the Academy being extended beyond the two Universities. At first the Fellowship resisted this radical idea; then the Captain offered to sponsor their first post war Retreat, in Cannes. Resistance collapsed and the Academy's Charter was modified to allow up to fifty 'Publishers of Good Repute' to become Fellows. From that point the Newtonian Academy became modern, began to refer to itself as 'NASP', and established its Annual Retreat as one of the highlights of the publishing year.

And here he was, Henri de Klompenmaker, invited to address this select group. He had heard of these annual retreats, where only the highest publishing strategy was discussed in the most rarefied of surroundings. They were, however, better known for the *luxe* of their locations and lavishness of the catering arrangements, than they were for the results of their deliberations, which were by tradition shrouded in secrecy. Discussions were meant to be 'off the record'. Nothing was published in the *Proceedings* and information exchanged was not to be circulated beyond the conference participants. The social aspects, on the other hand, were not only freely discussed, but were legendary.

Last year's retreat had been in Vail, Colorado (the tradition was to alternate between a ski resort and a golf resort) and next year's, which he already had hopes to attend, would be in Gstaad, Switzerland. Klomp had heard the tales of the banquet in Vail, 'like one of Nero's more lavish bacchanalias', and had seen the photographic evidence, including the grainy newspaper print of Jack

Hunter (Chairman, EuroScience Publishers) sitting naked in an open-topped sports car at the bottom of the beginner's piste. Only a solemn promise never to enter the state of Colorado again had kept him out of jail. Klomp shuddered involuntarily as he thought of the exertions the forthcoming week might bring. He felt his brain was up to it, but was his digestive system?

Thus pre-occupied, his steps took him to The Pavilion, which housed both the breakfast and the conference registration desk. As he drew closer, he heard raised voices and what sounded like tap dancing. He proceeded with caution. For Henri de Klompenmaker, discretion was very much the better part of valour. This exchange sounded distinctly combative and he took a peek around the door before entering. He was astonished to see Lionel Grove. What was he doing here? Klomp thought that the honour of representing SI at NASP had fallen solely to him. The other party to the dispute was doing the tap dancing, or rather stamping her feet repeatedly and loudly, as if to reinforce the point she was very vociferously making.

'Let me tell you, Mr Lionel Grove, when Lola Santiago is the Plenary Speaker, she expect to be met by limo at the airport'. Double stamp of feet. 'For this crummy meeting I have to take a cab. A cab!' Repeated stamping of feet.

'Dr Santiago. I have already told you that, apart from the Governor's Rolls-Royce, there are no limos in Bermuda. Everybody has to take a cab or a hotel shuttle'.

'Shuttle! Shuttle!' Continuous and rapid stamping of feet. 'What you mean, shuttle? When I consult for Bill Gates, does he tell me to take shuttle? When I consult for Rupert Murdoch, does he tell me to take shuttle? Lola Santiago does not take shuttle. She take limo, or she take next plane home'.

Lionel Grove was losing patience with Dr Lola Santiago. He didn't put up with this sort of behaviour from Donatella Versace and refused to countenance it in this woman. 'The next plane home isn't until Wednesday, so you'll just have to suffer Bermuda's hardships along with the rest of us. Now, I've got a busy morning ahead. Is there anything else I can help you with?'

'Help? Help? You have been no help. In fact you've been…'

'Dr Santiago. Save your breath for your presentation. There is nothing I can do about transport in Bermuda, but I can assure you that when you come to speak at the SI Strategy Meeting in May, I will personally arrange for you to be met with a limo'.

'You arrange that meeting too?'

'Of course. I am the SI Director of Corporate Hospitality. I also arrange the payment of the speakers and the media coverage'.

Dr Lola Santiago stamped her feet no more. 'You do? Media coverage?'

'Yes. I do. I am also the SI Director of Corporate Communications as well as Director of Honoraria'. It was almost 30 years since Dr Lola Santiago had left her native Cuba for the Harvard Business School. She had developed theories of pricing that were a bedrock of modern business and for which she had won the Nobel Prize. She had served on the Boards of Microsoft, Exxon, Disney and McDonalds. She now held the Chair of Pricing at the London School of Economics. She thought that she had encountered every known corporate species, but this was her first encounter with a Director of Honoraria. She was impressed..

'I guess you're a big shot at SI'. Dr Lola Santiago was invariably respectful to big shots, i.e. those who paid out large sums of money and controlled media coverage.

'Ralph Lauren seems to think so, but I'm just doing my job. I think you'll find everything ready for your presentation. You are the opening speaker, as you know. Also, I have arranged the personal scuba driving instructor, as you requested. I checked him out myself: he is very fit and his name is Sven. Will there be anything else?' Lionel Grove had shown Dr Lola Santiago that he was more than a match for her in the art of name dropping. She stamped her feet no more

'No, no, that all sounds great, just great. Thank you so much. I have my breakfast now'.

Rebellion quashed, Lionel proffered an olive branch. 'Nice outfit, by the way. Zandra Rhodes?'

'Yes, actually'.

'Isn't she a darling? So easy to deal with, unlike Some People'.

Dr Lola Santiago was, by now, feeling outgunned, and had no more names to drop.

Lionel Grove watched as Lola Santiago made her way to the breakfast room. She must be at least 50, he mused, but she could pass for 35. Her figure was impressively svelte: suspiciously so, in fact. Also, that chin, those eyes. If anybody recognised the tell-tale signs of cosmetic surgery it was Lionel Grove. An aficionado himself, he recognised a fellow warrior in the battle against the depredations of Mr Gravity. She certainly had style. The Zandra Rhodes blouse was not what he would have chosen to wear quite so early in the day, but Lola Santiago managed to carry it off.' Bold, very bold' he muttered to himself. Henri de Klompenmaker would have agreed. Awestruck by this titanic clash, he continued to lurk outside the door. Lionel Grove, who missed nothing, saw him and a bright 'Good

Morning, Henri', took him by surprise.

'Oh, good morning, Lionel. Was that the legendary Lola Santiago?'

'Yes, a most unpleasant and demanding woman'.

'I know, I mean, so I've heard. But she is the world's leading expert on pricing, so it's wonderful she could make it. She's great'. Klomp had read every one of her books.

'She certainly knows all about pricing. You know what her consultancy fee is? Twenty thousand dollars per day! Henry Kissinger would have been cheaper. I hope she's worth it'.

'She is, Lionel, she is. Nobel Prize, you know'.

'I know, and she's the rudest Nobel Laureate I have ever met'. Lionel Grove prided himself on his wide acquaintance with the Great and Good.

'But it's a wonderful catch for NASP'.

'I'm not concerned about NASP. If they want to throw their money away, it's up to them. The Chairman's invited her to speak at our strategy meeting in May. Two days at $20,000 a day. There's no budget for it. How am I meant to find that sort of money at the drop of a hat. And that's not all. She's flying herself down to Nice and wants us to pay for the fuel for her plane. I told her that's not an allowable expense. She said she'd take it up with the Chairman'.

'Who will say 'yes''.

'I expect he will. He's thrilled she's agreed to speak'.

'No wonder, she's *the* pricing guru. I've read all her books'.

'How very interesting. It seems to me that we have managed very well without her 'til now'.

Klomp, who knew that it was risky to dispute with Lionel Grove, thought it prudent to change the subject to something less controversial than Lola Santiago.

'A nice surprise to see you, Lionel, what are you doing here?'

'Organizing, as usual. The major publishers take it in turn to do this retreat and it's SI this year, which means me, of course. I'm Director of Industry Relations'.

'Good, things should go smoothly, then'.

'Naturally'.

'I heard last year was a shambles'.

'Shambles? Not at all, I was there for a day to discuss arrangements for this year with the Committee. Everything went like clockwork'.

'What about the Jack Hunter incident?'

'Oh that! Something like that happens every year. It was all dealt with very professionally. Nobody went to jail, after all'.

'Every year! What can they get up to in Bermuda? You can't even hire a car here'.

'You'd be surprised. But don't worry, I know the Governor. My father was his commanding officer during the War. Sir Donald will see us all right. Be sure to come to the reception at Government House on Tuesday, the Governor's Guard is Beating the Retreat specially for us. Wonderful uniforms'.

Much as he was in awe of Lionel's organizational skills, Klomp found some of the events he arranged completely incomprehensible. 'What on earth was 'Beating the Retreat'? A thought which he dared not share with Lionel, who would assume that he would Know About Such Things.

'Where is everyone?' he asked, to avoid obliging Lionel to give him yet another lesson on the more arcane British social customs.

'You're very early. I expect they will be swimming or playing tennis. They will probably turn up for breakfast around 9-ish. The opening session starts at 10, if Dr Lola Santiago can get herself ready by then. Have you booked your afternoon events?'

'What? Aren't there conference sessions in the afternoon?'

'No. The afternoons are traditionally kept free for informal interactions and brainstorming'.

'What does that involve?'

'I don't know exactly. I've never taken part in them. I use the time to go through my faxes, catch up on work with my assistant, call the office and check the arrangements for the evening. I don't have time for informal interactions myself, at least not here, but I'm told they are very productive. If you look at this leaflet, it shows you the different options. The deep-sea fishing is highly recommended, so I'd book now. My personal assistant will be along shortly and he'll be able to take care of it for you'.

Klomp was frustrated that he had not had time to digest the full programme, and wandered off to consider his options more fully. He hated not having things properly planned. Lionel Grove shook his head as he watched Klomp make his way to a chair in the corner. He hesitated to criticise The Chairman, but was surprised that he had sent Klomp to this retreat as his substitute. 'Hopeless', he thought. 'A back room boy, if ever I saw one. With his legs, he shouldn't be wearing shorts. What will the others think?' Actually, he already knew what Marlene Pym thought, and suspected that these thoughts were now being shared with the others at the beach during their pre-breakfast swim. As it would be a good thirty minutes before anybody else turned up to register, and everything was ready here, Lionel took himself off to the resort manager's office to check on the dedicated conference fax line, which had not been set up yesterday, contrary to his clear instructions.

Klomp studied the Afternoon Programme. Deep-sea fishing did not appeal and nor could he see how there would be much opportunity for 'informal interaction' while trying to haul in a barracuda. He scanned the list of options. Golfing was out, he didn't golf. Tennis? Croquet? Poker? No. A cruise in a glass-bottomed boat was more to his liking, and would probably provide a better opportunity for 'informal interaction'. That took care of one afternoon, but what about the other three? 'Historic Bermuda' sounded preferable to the 'Bermuda Perfumery'. The 'Crystal Caves' appealed; he had studied geology at university. Despite the fact that he had attained only a Master's degree, he felt an affinity for rocks. He would find out from the others at breakfast what they were doing. He would, of course, want to save one afternoon for a visit to the Post Office. Bermudian stamps were famously exotic.

He turned to the non-social aspects of the programme. He was honoured to be speaking at this evening's Plenary Session, chaired by the President of NASP himself. It was devoted to that old chestnut, 'The Future of Journal Publishing', about which many in the field seemed to be unnecessarily gloomy. He thought they were too sensitive to growing customer hostility and falling circulations. Neither of these need be an obstacle to increasing profits and keeping the shareholders happy, as his own talk would demonstrate. He, Henri de Klompenmaker, would show them how, with modern financial management techniques, journals could develop in a most satisfactory way, both editorially and financially.

Chapter 3: Another Journey

Although Fiona Hamilton PhD was by habit an early riser, 5.30am on this dark, cold March morning was a challenge even to her. But duty called and she wanted to be at Heathrow Airport in good time for her flight. She was travelling economy class on Air Lenin, an airline as notorious for its overbooking of flights as it was for the unreliability of its schedules. She had been urged to arrive early. Fiona was on her way to represent SI at the International Congress on Moleculetics, which was being held this year in Tallinn, Estonia. Her boss, Henri de Klompenmaker, had approved her journey only with the greatest reluctance, and was not convinced that attending the leading conference in the field, involving as it did lengthy meetings with many world-renowned experts, was an adequate substitute for the deskbound, numerical approach to publishing that he had espoused. But Tallinn was not Tokyo. The trip would be a cheap one, and hardly worth a lengthy argument. In fact, he rather relished the thought of Fiona Hamilton spending over a week in an hotel with unreliable plumbing and inedible food in an outpost of the collapsing Soviet empire. Perhaps it would curb her appetite for travel in the future…

She had joined SI after completing a PhD in chemistry at

Edinburgh University, and had made an immediate impact. Her combination of striking good looks and intellect, unusual in this, unglamorous corner of the publishing world, had brought a number of very successful publishing projects to the company, including the triumphant *Scripta Nanochimica Sinica,* and she had already been promoted twice. Her flair for publishing was matched by a love of it. No silver lining is without a cloud, however, and hers came in the form of the *Transactions in Moleculetics*, the reason for her trip to Estonia. The Editorial Board Meeting of the journal took place, by tradition, during The International Congress on Moleculetics. This was held every two years, in different locations around the world. Fiona's heart sank as she thought of the difficult discussions ahead. A purge of the journal's Editorial Board was, in her view, needed. She knew how second rate many of them were and that the main purpose of their trip was a fee junket at the expense of the publisher. She also knew that they were all cronies of Les Fyfe, the Editor *ad interim*, and that he would fight to keep them. Second rate himself, he had no intention of being outshone. What was worse, he was using the new numerical measures instigated by Henri de Klompenmaker to show how well the journal was doing. Yes, revenues were up; yes, profits were up; yes the number of papers published in the journal were up; and yes, these papers were, for some reason, now being much more widely cited. But she knew something was fundamentally wrong. As *Transactions* grew in stature within SI, it fell to new lows in prestige among the scientific community it was meant to serve. This convinced Fiona Hamilton of the fundamental flaws in the Klompean method. His cuts in staff and travel had weakened the links between SI and the many scientific communities it served. Nobody, apart from her, was actually going out and talking to them. Unwilling to incur the wrath of Klomp, her colleagues rarely strayed from their desks and relied on computer spreadsheets to tell them what was going on in the marketplace.

While she worried about the *Transactions* Editorial Board Meeting itself, Fiona knew that she would not lose sleep over the various dinners, entertainments and other diversions that had been arranged for them in Estonia by Lionel Grove, SI's Director of Corporate Hospitality. The Soviet Union may be collapsing, power supplies may be failing, the population may be living on watered-down cabbage soup, but Lionel would have ensured that what meagre resources remained would be fully deployed to ensure that SI met its hospitality obligations to its journal editors. Of this, at least, Fiona could be assured.

Arriving at Heathrow, Fiona, laden like a packhorse, struggled to the check-in where she was relieved of her heaviest bags and could progress to the departure lounge. There it was she saw Sir Henry Wiseman, FRS, Nobel Prize-winner and Master of St Leonard's College, Oxford. She did not know him personally, but had seen him on many occasions. He was the father of the field and had been a founding editor of its leading journal, *Acta Moleculetica*, published by Albany Press, SI's arch- rival. Of course he would be heading for Estonia. He had been one of the instigators of this international congress in the 1950s, and though now a septuagenarian, he was still a force to be reckoned with. Fiona was more surprised to see that he was accompanied by his formidable wife, Alberta, Lady Wiseman. Known universally as 'Bertie', she was a legend in Oxford and beyond. Lady Wiseman was as much at ease on radio panel games as she was on Government Committees and her plummy tones rang around the airwaves on a regular basis. There was no subject on which Her Ladyship did not have an opinion. She had been part of the Oxford scene for as long as anyone could remember. Her father was a very distinguished Dean of Christ Church, and she had graduated from Somerville in physics in 1931. She and Henry had married in 1938, but had spent much of the war apart. He in Naval Intelligence, she as a special assistant to 'The Prof', Lord Cherwell, who had been Churchill's scientific advisor. At that time, she had become accustomed to dining at Chequers, drinking good champagne, working odd hours, taking unforeseen trips at short notice, and being listened to. She had not lost these tastes, as the world found out from her autobiography '*A Dean's Daughter*'. Devoted to Sir Henry she would not, in view of his advanced age, allow him to travel overseas without her. (While her husband, at 79, was to her way of thinking, on the verge of decrepitude, Lady Wiseman, one year older, was, according to her, still in the prime of life. This she attributed to the beneficial effects of her daily trampoline session.)

Fiona found a suitable seat, arranged her packages around her and settled down to read her book. Time passed and the departure lounge began to fill. She vaguely recognised a number of faces, and it was clear that many of the passengers were destined for the Congress. The first announcement of a delayed departure took nobody by surprise. The inbound plane had been late in arriving and departure had been rescheduled for 12.00 noon. By now there were few seats left in the lounge, and at this point a stir at the door distracted Fiona from her book. She looked up and her heart sank. It was Les Fyfe accompanied by a rotund little woman, his wife Vera. They were a well-matched pair in height, circumference and in their volume of irritated chatter. Fyfe had been involved in *Transactions in Moleculetics* since it was launched. A scientist of no distinction, he had originally been assistant editor to the great Sir Edmund Jackson. When Sir Ed had disappeared on that trip to the Amazon, Fyfe had taken over *ad interim.* Sir Ed had been a great delegator, especially during the decade before his disappearance, when his interest in moleculetics had diminished and his interest in the preservation of the rain forest had increased. The retiral of Sir Algernon Brogue, who had always taken a close interest in the journal, allowed Fyfe to strengthen his position further. His grasp of moleculetics, never strong, did not grow, but he had become intimate with *Fowler's Modern English Usage*, which he had used as a weapon against unsuspecting and largely innocent non-English speaking authors. Thus was born science's first linguistic terrorist. A misused semi-colon was shot down with zeal, a split infinitive blown apart without warning. While the language barriers to publication in *Transactions* rose, the scientific barriers fell. The result was an author pool consisting of mediocrities from obscure British and American institutions whose prose style conformed to Fyfe's taste. Few reputable scientists now even bothered to read *Transactions in Moleculetics* and its issues lay, undisturbed, on library shelves throughout the world.

'Well, if it isn't Fiona Hamilton!' The broad vowels, a hybrid of Coatbridge (the Scottish town where he was born) and Rochdale (whose university faculty he had graced for 25 years), bellowed from two feet away, caused her to shudder involuntarily.

'Les, how nice to see you', cooed Fiona.

'Are those two seats free?' Fyfe occupied one without waiting for a response.

'I believe so'.

'Vera, Vera! There are two free over here'.

The little fat lady bustled over to the little fat man. Fiona, a devotee of classical physics, subscribed to the view that two objects could not occupy the same position in time and space, and that to defy this principle, as the Fyfes seemed about to attempt, would lead to a catastrophic event with unpredictable consequences. Two large Fyfes into two small seats would not go. But attempt it they did, and after some considerable disequilibrium involving the entire row, it appeared to Fiona that they had achieved their objective. But the laws of physics would not be so readily flaunted, and an unfortunate passenger at the end of the row was displaced onto the floor.

'So, Fiona, is everything arranged for the meeting?

'Yes'.

'We're not expecting as good a turnout as we had last time in Sydney'.

'I wonder why?', mused Fiona, who knew exactly why.

'Still, that'll save SI some money. There's been a lot of upset about your plans for the journal'.

'Really?'

'Oh yes, the Board don't like the idea of a new American editor at all. Not at all. We're very happy with Joe Simpson over there'.

'But he's not even American. The journal IS meant to be International, and most of the best work in the field is being done in America'.

'Yes, but we are a British journal and always have been. Any road up, we get plenty papers from America'.

'All from one campus, which happens to be where Joe Simpson, one of your former students, is based', said Fiona, tartly. She had already argued this point with him, and hated the cosy, parochial mediocrity that Fyfe cultivated.

'Bear State University is a very prestigious place', responded Fyfe, unconvincingly. 'Any road up, my friend Mr de Klompenmaker seems happy with the way I manage the journal. He tells me the profits are growing at a very healthy rate'.

'Yes, but only because SI has increased the subscription price by 30% in each of the last two years', said Fiona. 'I still think we need a new American editor in one of the major universities, which Bear State is not'.

'Joe Simpson's office costs the journal nothing. Bear are delighted to fund it. Do you think Henri really wants the extra cost of another office in the States when we have such a good deal with Bear?' Les smiled triumphantly. He knew that Henri de Klompenmaker would not countenance funding a second editorial office in the USA when they already had one for free.

Seeing that Fiona had no answer to this, or at least wasn't willing to offer one, Fyfe changed the subject. 'How about the Board Dinner arrangements and the plans for the reception we are hosting in Tallinn?'

'All arranged by Lionel Grove'.

That silenced him. One person whom Fyfe treated with great respect was Lionel Grove; a man of legendary organizational skills. 'Ask Lionel' was the cry of last resort at SI, whenever disaster loomed, which it frequently did. Had Napoleon been able to avail himself of Lionel's services his retreat from Moscow might still have happened, but the catering arrangements would have been better. Compared with the many challenges Lionel had confronted in the modern publishing world, the Russian Winter is a minor inconvenience. In recognition of the breadth of his talents, the Board of SI had showered Lionel Grove with an unparalleled range of directorships. From lowly beginnings as the personal assistant to the Editor of *Soigné*, Brogue & Co's leading fashion magazine, he had made spectacular progress through the ranks and, after a twenty-five year career, now held an unchallenged position of power at the top of the SI organization.

Lionel Grove's organizational skills were genetically programmed, his father having been the General in charge of ordnance for the British Army. Beyond the exotic uniforms of the cavalry regiments, however, young Lionel had shown little interest in things military. His remarkable interpersonal skills had been honed on the limp wrists and hissy fits of the fashion world of the 1960s. Sir Algernon Brogue liked the profits his fashion magazines generated, but had little empathy with the people involved. Nobody could handle Yves St Laurent as well as Lionel. Hubert de Givenchy was putty in his hands. Chanel had tried to steal him away from SI: but even Lionel preferred to keep a safe distance from the febrile atmosphere of the Paris *ateliers*. Promoted to Hospitality Director of *Soigné* in the 1970s, his events were legendary, his venues exotic enough to titillate the jaded palates of fashion's *haute monde*. By the 1980s, Lionel Grove's writ ran well beyond *Soigné* and encompassed all major SI corporate events. If the Queen was launching something for the SI Shipbuilding Division, Lionel made the arrangements. If the President was to make a top secret visit to an SI Defence Systems facility, Lionel personally supervised the lunch. He was now considered indispensable, not only because of what he did, but also because of what he knew. Over the years he had been witness to many bibulous lunches, dinners and other affairs involving senior SI executives and their distinguished guests. Not that he would dream of revealing a thing, of course: Lionel Grove is nothing if not discrete. But he did keep a diary, and they knew it.

Lionel Grove's involvement in events at Tallinn was in his capacity as Director of SI Corporate Hospitality. He had been involved in the arrangements for these meetings since Sir Ed's day, and had made it clear when Fyfe took over *ad interim* that he would continue to do so. Originally, because he spoke for Lady Jackson in a 'That is how Ed would have wanted it' kind of way, but now he did so on his own authority. Fyfe's one attempt to usurp this authority had been a disaster. He could still not recall the 1978 Congress in Manchester without a shudder. As it was on his own ground, so to speak, he had decided that Vera should make all the social arrangements, as it would put her at the centre of things and get her off his back. Never again. VIPs from around the world wandered around the airport seeking limos that never turned up. The restaurant where they planned to hold the Board Dinner had been double booked. The Ladies Outing to Chatsworth had been arranged for a day when the house was closed. And the Big Event – the 'Evening of Song with Gracie Fields' had completely bemused an international audience less familiar with 'Our Gracie' than the local population. Though the impersonator was in appearance the double of the great entertainer, this meant little to an audience who knew nothing of the original. The fact that she could not sing was more of a problem. Since then, in the small world of moleculetics, a 'Gracie Fields evening' had taken on a special meaning, often mentioned in after dinner speeches, and it was a meaning that Les Fyfe did not like.

There was another, much more important reason for Fyfe's deference to Lionel Grove. As SI's Director of Honoraria, he controlled the system that paid the fees to SI's journal editors. Lionel would know, but had never mentioned, the fact that Fyfe's payments were made into a Swiss Bank account, and would assume, but had never asked, that he would declare this to Her Majesty's Inspector of Taxes. That was one hornet's nest he did not wish to stir up. Yes, obsequious gratitude was his only option towards Lionel Grove.

'You know Vera, of course'.

'Yes, nice to see you again', Fiona lied.

'A pleasure', interjected Vera, who added, 'Ever been to Estonia? We had such a lovely time in Sydney. So much to see, although I did not get over my jet lag all week, my legs stayed swollen and I was a martyr to constipation. I hear the food is awful in Estonia'.

Fiona, who had heard the same, had come well stocked with nuts, chocolate bars and dried fruit.

'I'm sure Lionel will have seen us all right on the food front. He's never let us down yet', asserted Fyfe.

'I had no idea what clothes to bring for Estonia in March', whined Vera.

'I was told Wellington boots and warm clothes', suggested Fiona, 'It's cold and muddy'.

'If they insist on holding this congress in March it should be somewhere hot. I don't know what us wives are going to do all day'.

'We have to come to Eastern Europe every so often. You know that Vera. We need to show willing', said Fyfe, the diplomatist.

'You could have stayed at home', murmured Fiona under her breath.

Further low level noises, a mix of whining and digestive rumblings, continued to emit from Vera until suddenly she exploded with a 'Bloody hell, it's him', pointing towards the unmistakable, distinguished figure of Sir Henry.

'And her', added her husband.

'Haven't they given up yet'?

'He's receiving the Congress Gold Medal for Outstanding Contributions to Research. The first one', said Fyfe, almost choking on his words, 'Has to be there'.

Sir Henry and Lady Wiseman had been an uncomfortable fact of life for the Fyfes all of their married lives. In the great saga of thwarted ambition that this encompassed, the Wisemans were cast in the role of he- and she-devil. Demonisation began when Sir Henry had declined to offer Les a graduate student position in his laboratory in the 1950s; his second class honours degree was deemed by Oxford to provide insufficient evidence of the first class mind that Les was convinced he possessed. Never one to let such slights hold him back, he had made his way the colonies, whence he returned several years later with a PhD, to take up a lectureship at the Arkwright College of Technology, now the University of Rochdale. He had subsequently applied for positions at Oxford, Cambridge and St Andrews, and had been turned down every time. Wiseman had been on the selection committee in each case. Rochdale had thus become his reluctant academic home for over two decades. There was another source of friction. During Wiseman's long tenure as Editor of *Acta Moleculetica*, the original and leading journal in the field, every one of the papers that Fyfe had submitted for publication was rejected. The words 'Insufficiently original', 'Adds nothing new' and 'Dubious conclusions' in the rejection letters still stung. His active lobbying to be invited to join the Editorial Board of *Acta* had met with an equal lack of success.

Nor, he suspected, was the malign Wiseman influence confined to the academic world. Over the years, Fyfe had conducted a concerted campaign to become a consultant for a major pharmaceutical company – any major pharmaceutical company. He had heard the tales of the lucrative fees, the expenses-paid trips to meetings in the smartest resorts, and the bottomless well of research funding. Then there was the greatest prize of all, the Non-Executive Directorship! This provided all of the above, plus the Director's Share Option Certificate, which gave the right to buy the company's shares at preferential prices. Millions could be made from these deals. His honorarium as Editor *ad interim* of *Transactions* was chicken feed in comparison. Yes, it had funded the caravan in which he and Vera liked to tour the Lake District at weekends, but his heart was set on an air-conditioned Winnebago motor home, and that would require considerably more funds.

And who dominated the consultancy world and the boards of the pharmaceutical companies? Wiseman and his former students! Wiseman himself had done particularly well. His *Who's Who* entry listed at least ten non-executive directorships. What did he do with it all? Vera had always said that the Foundation he had set up to provide scholarships for students from Africa to study at Oxford was some kind of tax dodge to do with his share options.

What had Les got to show for his efforts? Fifteen years as a consultant to 'Pennine Nutritional Supplements', giving them ideas for making pigs gain more weight, chickens lay more eggs and cows give more milk. And on his appointment as a non-executive director last year, his only perks were a free supply of 'Laymore' for Vera's hens and thousands of worthless share options. Agriculture being in the state it was company profits had not risen in years. Fyfe was convinced that Wiseman had blocked his entry into the world of big-time corporate consulting just as effectively as he had blocked his exit from the University of Rochdale.

It was this sense of exclusion that had driven Les Fyfe to suggest to Brogue & Co, back in 1969, that they launch a new journal to compete with Sir Henry's *Acta Moleculetica*. There were enough rejected authors, like himself, to support a rival publication. He was stung that Sir Algernon Brogue had not considered him sufficiently distinguished to be the Editor of the *Transactions in Moleculetics*, but he was deemed fit to be assistant to Sir Edmund Jackson, whose own rivalry with Sir Henry went back to their days together at Oxford, when they were rivals for the favours of Alberta Gascoigne, the Dean's daughter. Co-founders of the field of moleculetics, their early disputes had been legendary. Eventually an unspoken accommodation was arrived at whereby they inhabited quite separate academic universes, to the extent that even their collections of honorary degrees were mutually exclusive. Those institutions that had honoured the one knew that a similar offer to the other would be met with a blunt rejection. An impartial observer might have judged that Sir Henry had the more distinguished academic career, but Sir Ed had registered more patents, had started his own company, and ended up the richer. Sir Ed had agreed to become the figurehead for the new journal only to annoy Sir Henry, which it did.

During the early years of Jackson's editorship, Fyfe's role in the journal had been in the boiler room rather than on the bridge. He processed the papers, corresponded with authors and wrote the rejection or acceptance letters as instructed by Sir Ed. The Fyfes enjoyed the travel and entertainment at the expense of Brogue & Co, but the financial rewards were negligible until, at last, Dame Fortune cast her golden smile upon them. Such was Sir Ed's passion for saving the Amazonian rain forest that a large part of his considerable fortune was devoted to this worthy cause. It had earned him a knighthood and the admiration of environmentalists worldwide. This admiration was not, however, shared by those who were keen to turn the rain forest into cash, which included SI's Division of Natural Resources, and he had gone missing on one of his regular expeditions to South America. A body had never been found and Oxford High Tables debate to this day, whether he was eaten by piranha or by tribesmen. Sir Henry favoured the tribesmen, as he was sure that the piranha would have recognised a kindred spirit. Whether aided by fish or by savages, Les moved quickly to consolidate his position. As soon as Sir Ed was reported missing, he took over '*ad interim*' as Editor. This interim position had now lasted for five years and *Transactions in Moleculetics* no longer met most of the objectives one might normally set for a scholarly publication. In the years since Sir Ed's disappearance, serious scientists had ceased to read it or publish in it; and it had become one of those journals of last resort where the mediocrities of moleculetics published the otherwise unpublishable. It did, however, continue to provide steady revenues and profits for SI.

'Is there no escape', moaned Vera. 'They must both be at least 90!' She could never see Bertie Wiseman without a shudder. As with her husband, 'Manchester 1978' would forever be engraved on her heart, and Bertie's presence was a heartless reminder of her contribution to that fiasco. As Vera's 'Evening with Gracie Fields' had lurched from one disaster to the next, many of the audience took up the invitation of the impersonator to 'Wish me luck as you wave me goodbye' and removed themselves from the hall. Gracie herself refused to reappear after the interval and Bertie, an excellent pianist, had saved the event with an impromptu concert. Not since Myra Hess's lunchtime recitals during the London Blitz had one pianist done so much for the morale of so many. To cap it all, Bertie ended the evening by reminiscing about the entertainments that the Real Gracie had given for Churchill at Chequers during the war. Yes, Manchester 1978 was the fixed point around which the subsequent social careers of Vera and Bertie had rotated, and neither could escape.

It was now 11.30am. Another announcement on the tannoy: 'Air Lenin regrets to inform passengers that the departure of Flight 1569 to Tallinn will be further delayed due to a technical problem with the aircraft. The estimated departure time is now 13.30'. A collective groan arose from the overcrowded lounge. Fiona sighed inwardly; 'another 90 minutes of the Fyfes'. Her mind turned to some possible escape plans. She could go to the restaurant for some coffee. No, the Fyfes would follow her there, and expect her to pay. What about the bookshop? Nowhere to sit and she would lose her place here. The ladies? Not with all those bags. Then came salvation.

'Lets go to business class lounge', Fyfe suggested.

''I'm flying economy'.

'Nay, lass, surely you can fly business'.

'No, I can't. Henri insists that non-directors take the cheapest possible fare'.

'Let's go', said Vera. 'I'm not sitting here on these so-called chairs for another hour. I need a drink. Up she rose and waddled off, expecting her husband to follow, which he did. Neither cast a backwards glance towards Fiona, nor did they invite her to join them in the business class lounge.

'Beer, doll?', asked Fyfe.

'Yes', replied Vera.

'Can you get me one too while you're up …'.

Muttering, Vera waddled over to the bar.

'I think that Fiona Hamilton is trouble', she opined.

'Who is she to tell you what to do? There's nowt wrong with Joe Simpson. He's doing a grand job at Bear State'.

'I can handle her. Nobody on the Board will entertain her half-baked schemes'.

'Still, as well to make sure the Board are on your side'.

'They will be, otherwise they're off; no more free travel, lunches or dinners. They know the score'.

'I suppose Bertie Wiseman will be coming on all the wives things. Bloody woman never gives up!'

'Look on the bright side, remember she used to work with Clemmie Churchill on the Russian Front during the war, so she might have some food parcels to share out. I think we might need them'.

'You know I never joke about food. The further she is away from me the better. 'Have you seen my *Woman's Own*?'

As his wife read, Fyfe contemplated his plans in the event of an insurrection by Fiona or Tom Carroll. Vera had highly developed trouble-spotting sensors and it was unwise to ignore her warnings. But he had two lines of defence, in which he had tremendous confidence. First, the craven support of the Editorial Board, which was packed with his cronies. Second, Henri de Klompenmaker, who would tolerate no interference with the *Transactions* moneymaking machine. As Les Fyfe sat in the business class lounge, nursing his beer, he recalled their first meeting.

'Why does *Transactions* reject 80% of the papers submitted', asked Klomp, in that very quantitative way of his.

'Well, we're told that the journal can't afford to publish more pages. The price would have to increase and the libraries wouldn't like it', had been Fyfe's reply.

'Rubbish. Have you never heard of inelastic demand?

We can just increase the price: they have no choice but to buy this stuff", retorted Klomp, who had not forgotten this basic fact of economics, learned two decades before at the *Ecole Europeenne de Compatabilite*. 'Most of these papers will be published elsewhere anyway, we are simply feeding the competition. I see I shall have to explain to you the purpose of journal publishing'.

Though he had, at that point, less than three months' experience as a journal Publishing Director, Klomp had already a number of very firmly held views on the purpose of journal publishing. He had developed these, from his perspective on the SI Corporate Finance Task Force over a number of years, and as Nigel Archer MBA fully endorsed them, he could now give them a public airing. 'The point', he went on, 'is to allow the author to publish and the publisher to make money. It is silly to reject all these papers. If we accepted more, everybody would be happy. The authors would be happy, because they'd all get published and the publisher would be happy, because he makes more money.' At first, Fyfe did not understand this and wondered where quality control came in, but then Klomp explained how he would make the Editor happy too.

'From now on, our journal prices are going to calculated by a formula based on the number of papers published, at a rate of £5 per paper. That would make the price of *Transactions* this year, £1000'.

'But that's what it is already'. Fyfe took a close interest in the financial aspects of *Transactions*.

'Exactly, but if you reduced the rejection rate from 80% to, say, 20%, you'd be publishing 800 papers a year instead of 200'.

'So the price of *Transactions* would be £4000?'

'You see my point'.

'Yes, but you'd never get away with it. You can't just quadruple a price like that. Customers would cancel their subscriptions'.

'But I told you, this is a classic case of inelastic demand. The price has little effect'. 'In any event, we would not do this overnight. We need to increase the number of papers published gradually, in a planned way'.

'But that's a big increase in my editorial workload'.

'You always said that you put as much work into an paper you reject as into one you accept. There shouldn't be any increase in your workload, should there?'

Fyfe could not refute this, though his jowls shook as he made an effort to do so. Klomp continued, 'But that's not the point, you should share in the benefits of growth. You would be paid on the basis of the number of papers you accept for publication.

Now Fyfe saw the point very clearly. He was being bribed.

Nigel Archer had also seen the point. If Klomp's ideas were correct, this business was truly the perpetual money-making machine that had been sought by Midas and Rockefeller. 'Let's try it', he said, and since then, from the perspective of the SI finance department, it seemed to be working. Indeed, they had based their most recent five-year forecast on the assumption that the number of papers published in their journals would grow by 20% per annum, their prices would increase proportionately, as would the resulting revenues. Profit growth would be even greater, as Archer's 'more system, fewer people' philosophy would reduce costs. The City and Wall Street had drooled over the forecasts for revenue and profit growth announced by the new Chairman within six months of his appointment. 'SI Back on Track' had been the flattering headline in the *Wall*

Street Journal. The share price had responded accordingly.

Naturally, Fyfe had accepted the bribe, which was a fee of £50 per paper published for himself, and £20 for his secretary. There had been some obstacles along his road to riches. Fiona Hamilton for a start, but Henri de Klompenmaker had made sure that her power to interfere was strictly limited. Then there were the remaining members of the original Jacksonian Editorial Board, but they were quickly replaced by Fyfe's cronies, who would do as they were told. As a result, the percentage of papers rejected by *Transactions* fell from 80% to 60% within the first year and to 40% by the end of the second. Fyfe's target, consistent with the Klompean laws, was to reject only 20% of the 1000 papers submitted annually to the journal. This, he calculated, would increase his editorial stipend to a healthy £40,000 per annum, while his wife Vera, his secretary, would supplement their annual household income by £16,000 as journal secretary. The Winnebago motor home was getting closer.

As he sat in the business class lounge at Heathrow Airport, Professor Leslie Fyfe allowed himself a private smile as he pondered the steadily increasing quarterly editorial stipend cheques that swelled his Swiss Bank account. 'What the Inland Revenue doesn't know can't hurt them', he told himself.

Left to her own devices in the departure lounge at Heathrow, Fiona Hamilton thought that much the most pleasant thing to do was to sit tight. Always prepared, she had packed a couple of apples and she proceeded to munch on one. The next hour passed quickly, and at last the first boarding call went out, for Business Class passengers only, who emerged from their lounge to claim their priority. The Fyfes skulked along the far wall, evidently keeping a safe distance from their *betes noires,* who did not notice them. Surely the Wisemans would be travelling business class, but neither made any move. Perhaps they were a bit deaf. But no, they sprang up when, eventually, passengers in 'Economy, Rows 20 to 33' were invited to board. Like her, it seemed, they would be right at the back of the plane.

Fiona Hamilton made her way up the aisle, with little help from the stewardesses. Air Lenin was famed as much for the rudeness of its staff as for the vileness of its food. By the time she got to Row 32, where she was sitting, the overhead lockers were already overflowing. She had paused to look for space, when an elderly woman sprang to her assistance. It was Lady Wiseman, who was in the next seat, with Sir Henry safely tucked in by the window. 'If only people would make better use of the space everything would fit in!' The area around them was instantly transformed as Bertie Wiseman cajoled every able-bodied man into moving bags from ceiling to floor and between lockers. Slavic mutterings of protest were met with acerbic replies in fluent Russian that resulted in places being found for all Fiona's bags.

'You don't believe in travelling light, do you, dear', commented Bertie.

'Publishers never travel light, I'm afraid'.

'Publisher, eh? With whom?

Her answer 'SI' did not seem to go down too well with the Wisemans, who both recoiled from her. Fiona had become used to this response. Her company was not the most popular in the business.

'What's a nice girl like you doing with that bunch of crooks?' quizzed Her Ladyship

'Bertie!' hissed Sir Henry.

' Sorry, dear, but the way they treated those poor miners in The Congo, and then their behaviour during the War. I suppose you're too young to remember all that. How rude of us not to introduce ourselves. I'm Bertie Wiseman and this is my husband Henry'.

Bertie began her forensic examination of Fiona's life and career, but was interrupted by the events surrounding the takeoff of an Air Lenin flight. The stewardesses attempted to persuade Fiona to yield her aisle seat to a rather large, portly man, but Bertie's fluent Russian was again deployed to good effect. Fiona was sure that she heard something like 'As Marshall Stalin said to me'. Sir Henry raised his eyes to the ceiling, but no territory was ceded and she remained in her aisle seat.

Once they were airborne, Lady Wiseman reached into her copious handbag and retrieved a bottle of duty-free brandy. 'Air Lenin does not believe in complementary drinks in economy, so we brought our own. Care for a brandy?' Fiona gratefully accepted.

'I suppose you'll be going to ICOM. Have you been before?'

'Oh, yes, I've been at the last three'.

'Henry and I didn't make it to Atlanta last time. Did y'all enjoy yourselves', drawled Her Ladyship. 'Tell me, are you involved in that dreadful journal?'

'You mean *Transactions*? Yes, I'm the publisher'.

'Hear that Henry! Fiona's saddled with *"Transgressions"* –our little joke, dear'. Bertie knew that this would provoke a strong reaction, and she was not disappointed.

'Should have been strangled at birth', were the words that emerged from behind the newspaper.

'Not you dear, I'm sure', soothed Bertie, who knew that the possibility of his remarks being misunderstood would prompt the gallant Sir Henry to emerge from his newspaper.

'Of course not! I meant that so-called journal', Sir Henry was now engaged in the conversation.

He gave a full and frank assessment of the many shortcomings of the major publishers in general ('greedy, appalling standards, monopolistic') and of *Transactions* in particular ('totally unnecessary', 'third-rate journal with a third-rate editor', outrageously expensive'). He asked Fiona what she was going to about it. She told him, but chose her words carefully, as she felt she was in hostile territory.

'We need to review the composition of the Editorial Board'.

'Sack the lot of them. And the Editor *ad interim* too' was Sir Henry's considered advice.

'And the Editor's wife', added Bertie.

'We also need to improve the quality of the papers published'.

'Suggest to the Editor *ad interim* that he might consider rejecting the occasional paper'.

Fiona was painfully aware of the reputation the journal had gained for publishing anything.

'We need to reposition the journal to cover the more topical areas'.

'Impossible, when it is run by that crowd of charlatans. Jackson was by no means wonderful, but since Fyfe took over! I remember bumping into Algy Brogue at the memorial reception in the Athenaeum for Ed Jackson. I warned him about Fyfe. Then he retired on us and suddenly Brogue & Co became the Publishing Division of SI. Brogue was no Albany Press, of course, but at least it was run by a gentleman.'

'I understand your concerns', said Fiona, at her most diplomatic, 'but it is my job to try'.

'Good girl. You show 'em', shouted Bertie. 'Have another brandy'.

Fiona explained to Sir Henry how she had assembled a strong body of evidence that largely supported his negative opinion of *Transactions*, which she would present at the Board Meeting. He listened with some sympathy and even made the occasional constructive suggestion. Sir Henry Wiseman, recognizing a fellow-scholar, was softening towards Fiona Hamilton.

Chapter 4: Bear State University

The first-time visitor to the Todd Institute for Moleculetics at Bear State University could be forgiven for thinking that he had come to the wrong building as he entered the enormous shining pyramid and walked through a glass-walled tunnel, which was clearly under water. Above, were those the brightly clad outlines of surfers scudding back and forth across an evidently choppy indoor lake? And was that the sound of the Beach Boys in the background? Having emerged at the reception desk at the far end of the tunnel, the visitor would be informed that they had just entered the Franklin J Todd IV Institute of Windsurfing, that tours had to be pre-booked and that there were no more places for that day, a fact that would be evident from the long lines waiting to enter the different attractions, variously labelled 'Live windsurfing display', 'Theatre of Surfing' and 'Windsurfboards Through the Ages'. Those interested in moleculetics rather than windsurfing would be directed to the third floor 'Academic Reception' where a transparent floor might reveal the spectacular 'Formation Windsurfing Team' in action on the indoor lagoon below. Above the third floor, the building hummed with scientific instruments and was inhabited by the people in white coats who populate laboratories throughout the world. This was, however, California, and underneath the white coats they wore beach sandals and shorts, ready to hit the surf at a moments notice. Bear State University is what is known as a 'Party School', a grove of academe where prowess in the bar or on the beach confers greater prestige than more traditional measures of academic performance. Since it was founded in the 1920s, Bear's oceanfront location had guaranteed its pulling power for a certain type of student. Its other main advantage, at that time, was its distance from the fleshpots of Los Angeles, a strong recommendation for the weary, well-heeled parents of the more academically challenged of the Southern California elite. As a result, this small, private university had basked in sunshine and in the donations of grateful or grieving parents for many decades. Bear had one of the most attractive campuses in the nation, with manicured lawns, palm-lined avenues, lavish sports facilities and a collection of buildings designed by the Who's Who of American architecture of the 20th century. Bear came nowhere in the league of academic distinction, but came top in the league of student excess. Its undergraduates and alumni killed themselves at an alarming rate in car crashes (vehicle ownership was encouraged and Bear was the first university to initiate campus-wide valet parking); sporting accidents (usually in the surf on their private beach, skiing or climbing at the Bear Mountain campus, diving at the Bear Caribbean campus, or hunting big game on the Bear Africa campus); and of course, from the usual drug overdoses.

Every cloud has a sliver lining, and the resulting high mortality rate among students had been matched by an equally high donation rate from bereaved parents. Indeed, the Board of Governors had come to regard a regular flow of such donations as an actuarial certainty, to the extent that the monies anticipated were included in their strategic plans. When the inevitable tragedy occurred, the President, George T Winthrop VII, having checked the Fortune 400 list, would, if he sniffed a substantial donation, fly in person (in the University's private jet, a memorial to Thomas J Boeing III) to console the grieving parents and to offer them the opportunity to assuage their grief by funding a much-needed new sports facility, dormitory, or overseas outpost.

Bear did not have a major medical school for the same reason it did not have a major law school. It was felt that this might attract the wrong sort of student. There were, nevertheless, many doctors and lawyers among the parents of Bear students, and Winthrop's predecessor as President, the revered Theodore J Hughes VI, did not wish to set unnecessary obstacles in the way of their considerable donating power. He devised a strategy whereby Bear could continue to attract the right sort of money without attracting the wrong sort of student. The essence of this strategy was 'nichisme', which involved targeting specific fields within medicine and law that would not only attract the right sort of student, but which would also provide a service to Bear students, alumni and their moneyed relatives. The results had been The Bear School of Divorce Law, the Bear School of Cosmetic Surgery and the Bear School of Fiscal Studies. All three had established fine reputations, and provided services much appreciated by the student body and their families. Bear tuition fees may be the highest in the state, but the discount rates for nose-jobs, divorce settlements and tax avoidance schemes by the finest practitioners of these skills more than made up for this, and probably accounted for the large number of mature students at Bear. President Hughes had died (tragically, choking on an olive while downing a martini in the course of his annual pastoral visit to the Caribbean campus). Sadly, this left the plans for Bear School of Cryogenics and the Bear School of Orthodontics on hold. President Winthrop's development priorities lay elsewhere, specifically in the direction of a large, virgin tract of land on the northern fringes of the campus that he thought would be ideal for a new Bear School of Golf and championship golf course. This would be the monument to his reign.

The graduate school at Bear had always been small, and with the exception of the truly pioneering work in belly-tucks (The Bear Tuck) and the virtual invention of the concept of 'palimony' (The Bear Necessities), Bear was a name rarely seen in the leading scholarly journals. President Winthrop was, therefore, caught unprepared when he paid his consolatory visit to Mr and Mrs Franklin J Todd IV on the occasion of the untimely death of their son Franklin V, tragically trampled by buffalo, while studying at Bear Africa. The Todds were one of California's leading families, but were known to be wildly eccentric. The first Franklin Todd had been a '49er' and had built the family fortune during the gold rush; subsequent investment in railroads and property had consolidated their position. Recent generations had married for beauty rather than brains, and were, in consequence, stalwart supporters of Bear. Grief had not dimmed the Todd eccentricity and President Winthrop was shocked when they rejected his proposals to fund a golf school in memory of their son. The Bear Funding Office (motto: 'Where there's a Will there's a way') had assured him that all the Todds were avid golfers and he was certain that he would at last secure funding for his cherished Bear School of Golf.

He was even more shocked by the Todds' counter-proposal – a research laboratory. This was unheard of at

Bear. The President, an 'Old Ursuan' himself, pleaded with them to reconsider. He threw decades of cast iron tradition at them, 'Bear had NEVER been a research school'. Attracting large numbers of graduate research students to the campus would fundamentally change the much-treasured character of the place. 'Fill it with eggheads', was how Winthrop had put it. Research also involved hideously expensive outlays. More money would have to be spent on the library for a start. The present building had served its purpose more than adequately since the 1920s, and Winthrop was convinced that if it were not for the bar and the sundecks with ocean views insisted upon by its original benefactor, even fewer students would use it. The periodicals collection, currently dominated by *GQ, Field and Stream* and *Surf's Up*, would have to be supplemented with all these dry-as-dust research journals, and he had heard that these were ruinously expensive. He would have to talk them out of this idea.

But the Todds would not be dissuaded. A research laboratory it would have to be. Prunella Todd's father was a gentleman scholar, who had his own chemical laboratory at the family seat in Scotland. She was keen to fund something that he would admire. Determined to get them into a more sporting frame of mind, Winthrop had thrown the memory of their son at them. 'Surely they remembered that his lifelong passion was windsurfing'. Indeed it was. Since winning the California Junior Championships at the age of 8 he had been devoted to the sport and had led the Bear team, 'The Grizzlies', to memorable victories in the World Championships in the Caribbean two years previously.

'He has a point, Prune', said Franklin IV. 'And there's all that stuff', he added, recalling that their son had one of the most comprehensive collections of windsurfing memorabilia in existence. The conservatory at their Montecito mansion was packed with it. His collection of equipment traced the windsurf board from its earliest days as convenient personal transportation for the islanders of the South Pacific, to the latest models used at championship level. He had designed several pioneering boards himself, using the latest materials and the most aerodynamic forms. These, together with his collection of windsurfing accessories, videos, posters and trophies had been deposited with his parents when he had left for his sadly curtailed year at Bear Africa. Prunella had complained about her treasured conservatory being taken over in this way, but she could deny her 'Frankie' nothing. Now she saw her opportunity to reclaim it.

'Your right, I need the Orangery back for my parrots. I have had to suspend my breeding programmes, and they're plucking their feathers since they were moved into that small aviary'. She mused, 'OK, we'll fund a Museum of Windsurfing in memory of Frankie'.

This was one building project outline that Winthrop did not have in his portfolio, but in desperation he clutched at it. Anything was better than a research laboratory. If Bear was not one of the world's centres of excellence in any of the more traditional branches of scholarship, it was, so to speak, the 'Caltech of Windsurfing'. Increasing its profile on this front would also attract very much the 'right sort of student' – rich jocks and jockettes. Winthrop knew nothing about the sport, or indeed much about any sport but golf, but he could see that it was entirely consistent with Bear's motto *'Mens Pulcherra in Corpore Pulcherimmo'*. Perhaps they would fund a 'School' and 'Chair' also, the 'Todd Professor of Windsurfing' did have a certain ring. The more he thought about it, the more he liked it.

Prunella knew exactly what she wanted, 'Just like that museum in Lausanne; remember, the one with all these old skis, bicycles and running shoes next to the Beau Rivage'.

'The Olympic Museum?'

'Yes, that's what we want, but not so boring. I want lots of running water. 'Frankie loved the water. And wind'. Mrs Franklin Todd IV was now in full flood.

'I'm sure we could commission a spectacular building', enthused the President, as he mentally scrolled through the list of possible architects (he had got as far as 'P' for Pei)'. But don't you think that a fully-fledged Institute would be a more fitting memorial than a museum? You know, with faculty, courses, students and all that?'

President Winthrop thought he was home and dry, when Prunella said conclusively, 'Excellent idea! It's decided then. An institute of windsurfing, with a chemical research laboratory above'.

Winthrop croaked, 'A chemical research laboratory above?'

'Above, obviously', affirmed Mrs Todd, 'so that the smells don't stink out the windsurfers'.

'Yes', agreed Mr Todd.

'But we've never had a research facility at Bear'.

'You will now!' asserted Mrs Todd.

'What about something nice and big, like that thing up at Stanford?' suggested Mr Todd.

'We don't have room for a particle accelerator!' replied a very panicky Winthrop, thinking of the land earmarked for the championship golf course. 'And it would have to be underground'.

'No use, Franklin', agreed Mrs Todd, we're not paying for a hole in the ground. I want something chemical. You know my father loves doing experiments in his little lab in the old carriage house at Nethie. This might even persuade him to visit us in California', added Prunella, whose father hated to travel.

Winthrop had an inspiration. 'A drug abuse research Institute would be perfect. Our undergraduates could really get behind that'. Not a wise proposal, given the Todd family history of recreational substance abuse.

'Certainly not!' bellowed Mr Todd.

'Too sordid!' added Mrs Todd. 'Look here, you're the egghead, so think of something acceptable that I would not be embarrassed to show my friends. But it must have something to do with molecules. My father is fascinated by them; blew himself up once. Now, I've a session with my acupuncturist, so I am afraid we'll have to let you go'. Prunella stood up and gave Winthrop her hand.

Franklin IV silenced further protests with 'Budget no more than $100 million. Look at the Bermuda angle. And the tax write-off. We need to get something back from old Uncle Sam'. He turned to his lawyer, 'Ed, you know what we want. Sort something out with President Winthrop'.

'I'll be in touch to work through the details', said Ed, who had been silent up until now.

'But...but...' said the President, torn by conflicting emotions of greed and fear. Both Todds were now on their feet, smoothly thanking him for taking the trouble to come to console them, assuring him of their commitment to Bear and handing him over to Desmond, their butler, who escorted him to the waiting Todd limo. The gates swung open and the car purred through the wooded hills of Montecito towards Santa Barbara airport.

As the Bear jet made its way down the coast, the President was still trying to take in the Todd proposal. He contemplated the potential disaster facing him and downed the double scotch brought by the stewardess. 'A research laboratory', he groaned to Monica Parsons, his PA. 'They want to fund a research laboratory. Who'd have though that Bear would come to this! And on my watch! How will I explain it to the Board of Governors? I've told them we're sure to get the golf course this time. Nicklaus is already working on the design'.

'But I thought that all the Todds are keen golfers'.

'But not Prunella Todd, who seems to call the shots on this one. 'It seems her father is some kind of alchemist back in England. In desperation, I tried to get them onto windsurfing, which was the only thing young Frankie was ever good at'.

'But we already have the best surfing beach in California on our doorstep. Bear's already state-of-the-art on surfing. We don't need anything else there', asserted Monica.

'Oh yes we do. We need the Franklin Todd V Institute of Windsurfing. Prunella insists. Their lawyers are drawing up the documents as we speak'.

'What!' slurped Monica, through her mouthful of Chardonnay.

'I'm afraid so. The ground floor is to be the museum, with the laboratory above'.

The governors could just about live with the museum, so long as it is devoted to sport. I can see IM Pei'.

'So can I. Such a nice addition to the Frank Lloyd Wright, Julia Morgan and Buckminster Fuller stuff on campus'.

'But I don't think IM Pei does labs, at least I have never seen a glass pyramid lab'.

'What about that British guy, you know the one who does buildings with pipes sticking out all over the place. Ideal for a lab'.

'Rogers, I think. You have a point'.

'I had the strong impression that the museum/lab combo is non-negotiable. This is our first shot at Todd money.

They've been coming here for generations and so far, not a cent, if you exclude the fines for unacceptable behaviour'.

'It would certainly be a feather in your cap to be the first President to tap the Todd goldmine. We'll have to work on the Governors, though', added Monica. 'How much, by the way?'

'$100 million'.

'That's quite feather!' shrieked Monica.

'The biggest ever'.

'Where there's a Will, there's a way' – Monica reminded the President of the Bear fundraising office's motto. The both laughed, settled into their deep, leather chairs and continued the discussion in a more optimistic mood.

Monica pointed out that the Todds did not seem to be too fussy about the research field the laboratory would cover. 'Something to do with molecules' had been the only stipulation. That would give President Winthrop the opportunity to make some suggestions of his own, and direct them towards topics that would limit the damage to Bear's status and ambience. High-energy physics had, fortunately, already been ruled out by the Todds themselves. Anything too biological was also out of the question. Winthrop did not want the campus over-run by laboratory rats. As he pondered, Monica flicked through the latest issue of *The Economist*. Her eye fell upon a heading in the Science and Technology section: *'Moleculetics; the science of designer molecules'*. A fanatical devotee of Gucci, Hermes and Chanel, she read it avidly. It seemed that in this 'rapidly developing' field of science, the idea was to decide on the task you wanted done and then design a molecule that did it... Her thoughts turned to cosmetics, anti-wrinkle creams, hair tonics and other elixirs. An Institute devoted to designer molecules would certainly fit in with the Bear ethos, and was also appropriately 'molecular', which would keep Mrs Todd happy.

'I've got it', she cried, tossing *The Economist* to the President. 'Look, that's the stuff for your research lab'.

President Winthrop agreed, the Board of Governors agreed, and the Todds agreed. The architect was commissioned and the glass pyramid of the Franklin Todd V Memorial Institute of Windsurfing and Moleculetics arose, instantly becoming one of the great sights of the California coast. Each day saw long lines of students and visitors seeking the thrill of the ultimate windsurf experience. The Institute of Windsurfing was quickly operating at capacity. Such was its popularity that the trustees, with the full support of the Todds (whose entrepreneur genes had not been bred out of them entirely), decided to limit numbers by imposing an entrance fee, except on Wednesday afternoons, when the Friends of the Todd, who also happened to be Friends of Prunella, had free entry. Even with an entrance fee, however, the Institute continued to attract huge crowds and reservations for peak times had to be made months in advance.

Attracting research scientists to the Todd Institute for Moleculetics had proved rather more of a challenge. Calls from Bear were not taken too seriously in the academic world. The President's usual approach to filling faculty positions, searching the list of Old Ursuans, had drawn a complete blank. No Bear alumni had gone into scientific research. An executive recruitment agency was hired at enormous expense, only to be rudely rebuffed by the candidates they had identified at Harvard, Yale and Berkeley. 'Bear, are you nuts?'; 'I don't want to kill my career just yet'; 'Bear, I'd fail the drinking test'; 'Surely some mistake, I can't surf'.

For two years the Institute's state-of-the-art laboratory facilities remained empty. On the Todds' regular visits, they rarely ventured to the upper floors of the pyramid, but just in case, on those days, the no-expense-spared laboratory equipment hummed into life, peculiar odours filled the air and dozens of Bear students were dressed in white coats, promised some kegs of beer and told to 'look busy'. The Board of Governors was becoming uneasy, as the income from the enormous Todd Endowment and the ever-growing Windsurfing Institute surplus lay unspent. The local press and TV companies were beginning to sense something was amiss and were asking for interviews with the Director of the Institute for Moleculetics. In panic, the Board of Governors took the unprecedented step of appointing a Dean of Research. The shock waves reverberated around the campus and there was much hostile comment in the student newspaper. Competing claims for what undergraduate opinion regarded as an extravagant waste of money were put forward. What about the sports injuries clinic, the hot-tub technology centre, or the long-needed skateboard park? Any of these, it was felt, would have a better claim on funds than a Dean of Research. Demonstrations were held, and the Governors only prevented a major riot by giving the student body two specific assurances. First, that this was NOT the thin end of the wedge and that no more donations involving research facilities would be accepted. Second, the new Dean would be an Old Ursuan who would be sensitive to the culture of the campus.

The search for a Dean yielded one acceptable candidate, George Q. Huntingdon, Head of Alumni Relations at Rhode Island Episcopalian College. He had briefly strayed from the Bear norm by obtaining a post-graduate degree in geology, before deciding that research was not for him. Relieved to find an Old Ursuan with any post-graduate qualification, even a mere Master's degree, President Winthrop decided he would do. Huntingdon was called, accepted, arrived and commenced the task of populating the Todd laboratories. Still it did not prove easy and he too was rebuffed by the obvious candidates. He decided to cast his net wider, and now his post-graduate research experience, though short, began to show. He visited the Bear Library to search the journals. *GQ*, *People* and even *Time* failed to yield any names associated with moleculetics. Even *The Economist*, which had been the genesis of this whole idea, listed only those individuals who had already said 'no' very loudly and clearly. He had, in *The Economist*, however, seen a reference to another journal that might prove more fruitful. The reference was none too flattering, but it was his only lead. A copy of the *Transactions in Moleculetics* was sent for, arrived and was studied by the Dean of Research for the greater part of an hour. He decided to contact the Editor, who was in England, for advice. A fax was duly sent to Professor Leslie Fyfe at the University of Rochdale.

The response was immediate.

'Dear Mr Huntingdon

Thank you for your fax. You were right to seek my advice on this subject, as I have worked in the field of moleculetics for two decades and know every scientist of consequence. It is my mission to promote the growth and wellbeing of moleculetics and I would be delighted to assist Bear State University in its search for suitable personnel for the new research institute. I am not familiar with your university however, and would be grateful if you could send me a copy of your Prospectus.

As a first step, I propose that you set up a Search Committee. I would be willing to chair this. I can also provide you with a list of eminent candidates to join this Committee, which should meet in California as soon as possible. (I take it that Bear would pay travel expenses and I am confident that we can come to a satisfactory arrangement on my level of remuneration).

I look forward to your reply.

Yours sincerely

Leslie Fyfe PhD, MBE

Ridley Professor of Moleculetics'

The Dean was so relieved by this reply that he kissed the fax and danced a little jig. At last, a positive response. He was as ignorant of Rochdale as Fyfe had been of Bear. As a sailing man he had been to Cowes; as a golfing man he had been to St Andrews; as a rowing man he had been to Henley; as an American he had been to Wimbledon; but nothing had taken him to Rochdale. He would look it up in his atlas. Meanwhile, he would fax an invitation to Professor Fyfe to visit Bear at his earliest convenience. Again the response was instant and bade the Dean call the Professor. The following telephone exchange ensued:

'I'll have to bring my PA'. (i.e. Mrs Fyfe)

'Certainly'.

'Will you take care of a hotel reservation?'

'Bear has its own guest house overlooking the beach complete with butler, cook and chauffeur'.

'My PA may not need to be present at all our meetings. Can someone take care of her when we are busy?'

'The Bear concierge service will take care of her'.

'I take it that business class flights will be funded?'

'The Bear corporate jet will come to meet you in England. Which is your nearest airport?'

Fyfe could hardly contain himself.

Professor & Mrs Fyfe duly paid their visit to Bear. A Search Committee, composed entirely of Les's cronies, was set up, meetings were held, champagne was drunk, life was made very unpleasant for the butler, cook and chauffeur of the Bear Guest House. The Director and Faculty of the Todd Institute for Moleculetics were duly appointed. The laboratories filled with white-coated boffins. State-of-the-art machinery hummed expensively into action, exotic aromas filled the air, seminars were given, papers published, and there was a generally satisfactory air of activity about the place. Despite this, for the next five years the work of the Todd Institute went largely unnoticed in the wider scientific world.

Joe Simpson, graduate of the University of Rochdale, had been The Search Committee's choice as Director of the Todd Institute for Moleculetics Nobody could deny that he was diligent, or that he was a hard task master. He set an example to his colleagues by arriving at the laboratory at 7.30 every morning and rarely left before 7.00 in the evening: in fact, he had no life outside his work. Despite this industry, hardly a ripple of an original idea emerged from the Todd to stir the pool of existing scientific knowledge. No new avenue of enquiry was opened up. No brilliant flash of light was shed upon a hitherto intractable problem. Nothing close to a new discovery occurred. Yet, dozens of apparently well qualified and evidently highly paid people worked all hours under the watchful eye of their Director . Its laboratory facilities were the envy of many another college. Simpson was denied nothing, for fear of alienating the benefactors. Franklin and Prunella were delighted with their new toy, and shared this delight with their many friends during their regular visitations.

An entire Giza of glass pyramids sprouted around the Todd, funded by the healthy profits from the Windsurfing Institute and supplemented by the usual crop of untimely student demises. Just as the civilisation of ancient Egypt was based on the annual flood of the Nile, so the political economy of Bear was based on the equally predictable rate at which its reckless student and alumnus body met untimely ends. A major task of the Bear funding office was to scan the world's media for headlines such as 'American bird-watcher eaten by tiger', 'Leg of missing American surfer found in shark stomach', 'US diplomat killed in explosion-illicit laboratory facility suspected'. A statistically significant percentage of such deaths were found to be Old Ursuans, which made the subscriptions to the Associated Press and Reuters news services an excellent investment for Bear.

Even at Bear, the number of such deaths was finite, and frustratingly lower than the number of new projects that President Winthrop wished to endow. He had still not been able to find a sponsor for his cherished School of Golf. But, having tapped into the potentially rich vein of Todd funding, he had hopes to mine it further. His policy, for the first few years of the Todd Institute had been to humour Simpson and accede to his every demand. It was only when, at Simpson's prompting, he felt obliged to propose to the Board of Governors the building of a new library that eyebrows were raised. They had tolerated the pyramids hosting the new 'NMR facility', 'X-ray crystallography centre', and 'Mass spectrometry lab', but they felt strongly that a new library would fundamentally change the character of Bear in a way that they did not like. The student body was becoming restive again, as they felt that earlier promises about no further growth in the research school were being broken. A long-planned roller-blade arena remained unfunded. The expansion of Bear Caribbean was on hold. Resentment against the 'eggheads' had already grown, to the extent that they were asked not to wear their white coats outside the laboratory.

These arguments were in full flood when an Alpine avalanche resulted in the extinction, in his sophomore year, of an heir to the Blake publishing fortune and the consequent erection of a further glass pyramid to house the new library. The Blake family had no objection to pyramids, but they were insistent on a library – good for business. As students seldom visited the facility, the librarian could keep everything in perfect order. Books and journals rarely left the shelves and she and her staff were hardly disturbed. Initially, the scientists at the Todd, the original *raison d'etre* for the Blake, had been troublesome. In fact, they had become all-too-frequent users of the library, but the librarian had solved that problem by persuading them that it would be much better for them to have the journals and books they needed in a special reading room in the Todd itself.

Chapter 5: The Todd Institute Makes an Impact

Joe Simpson, Director of the Todd Institute for Moleculetics, had responded with his usual energy to Professor Leslie Fyfe's request for 'more papers for *Transactions* '. He had always enjoyed basking in Fyfe's approval, and 'the hardest working graduate student in Rochdale' had become the 'hardest working editor of *Transactions*'. His appointment as the journal's US Editor had not been met with the acclaim he might have expected. The number of invitations he received to present papers at major conferences had not increased. He had not, as yet, been invited to sit on any Committee or Board outside of Bear, and his appointment had not led to the immediate, expected increase in papers from American authors for *Transactions*. Simpson applied his considerable energies, as well as the considerable resources of the Todd Institute for Moleculetics, to correcting this infelicity. His strategy was two-pronged and was announced at his regular, weekly Group Meeting.

These were dreaded by all in the Todd, with the exception of Simpson himself, who looked forward to them mightily. He did not loom large in the wider world of academic research, but here he was king, and a well-funded king at that. In research, as in most fields of human endeavour, money talks, and when Simpson controls the purse strings, it bellows. By tradition, the Group Meeting was held on Friday evening, as this ensured that his staff would have some new problem to think about over the weekend. Of his four department heads, only one was American, and he was the only one who ever challenged Simpson. Tom Carroll was used to adversity. The son of a rancher on the central California coast, he had grown up in an environment where education was regarded only as an unwelcome distraction from helping his father run their herd of cattle. His interest in science had been sparked early in his high school career, and his teachers had been impressed by his aptitude for it. This cut little ice with his father, who allowed him to go to the local state college only grudgingly, insisted he continue to help on the ranch and was unimpressed with his *summa cum laude* degree in chemistry.

Tom was offered a place at some very distinguished graduate schools, but decided on Bear, where he could combine his passion for research with his passion for surfing. He was the first American graduate student to join the Todd Institute for Moleculetics. Tom was a star from the outset, completed his doctoral research in record time, and was duly awarded the first PhD in the history of Bear. The fact that he not only had his own ideas, but also questioned Simpson's had led to many conflicts in the early days of their relationship. But once it became clear that his papers were being accepted for publication in journals such as *Acta Moleculetica* and even *Nature*, and he was being invited to give papers at the leading conferences, an uneasy truce was reached. Simpson eventually, and with great reluctance, was compelled to appoint him to a tenured faculty position.

Tensions grew when Joe Simpson was appointed American Editor of *Transactions*. He expected his colleagues to publish in 'his' journal. Tom categorically refused to do so and expressed his opinion of *Transactions* in the most colourful language. The last Group Meeting had been enlivened by an open argument between the two of them on this topic. Simpson was, for once, dreading this Friday's event. There was to be an update from Tony de Lucca on the work that he and his students had been doing on the neurochemical basis of hunger. This had generated many papers for *Transactions* over the last two years which had, apparently, been very highly cited. Even more important, from Simpson's point of view, was Prunella Todd's excitement about this line of research during her last visit. Both Todds usually sat through the Institute's scientific presentations with a marked lack of interest, eager to move on to see the latest improvements in the windsurfing facility. This year, however, de Lucca's description of how a new class of compounds he had been developing could apparently suppress the appetite of rats, and the graphic evidence of it in a video clip of rats actually ignoring food had prompted shrieks of delight from Mrs Todd.

'At last!' she exclaimed, 'Something useful!' and after the presentation insisted on being taken by Joe Simpson on a tour of the laboratory rat facility to see the evidence for herself. It was impressive; the rows of cages on the right were filled with rats as slim as supermodels, their food bowls barely touched. Their constant, somewhat erratic, movement attributed by Prunella to a devotion to aerobic exercise. Such a contrast with the cages on the left, where sat the couch potato rats; fat, sedentary and somnolent, except for the occasional excursion to check whether the empty food bowl had been replenished. Simpson showed her the daily food allowance – exactly the same for both sets of rats, as well as the small amount of Compound X that was added to the food of the sylphs. Prunella had many questions. How often are they fed? How much of Compound X are they given? Are there no side-effects? Does it work on other animals? How long before it is available for humans? How much of this stuff do you have? To this barrage of questions, Simpson had ready and, to Prunella, convincing answers. His answer to her last question was most impressive of all. He unlocked the store room, where the shelves were filled with jar upon jar of Compound X. She gasped in awe. Simpson's beeper rang.

'Excuse me for one moment, Mrs Todd, I need to take phone call'.

Left alone, Prunella picked up a jar, and examined closely the white, crystalline substance inside.

'Looks just like sugar'.

She opened the jar and sniffed.

'No smell'.

Prunella Todd had a thought.

'They'd never miss a couple of jars. That dog of Franklin's could do with losing some weight: he's always giving her nibbles'.

Over the years, Prunella had argued with her mother, Lady Maitland, about most things, including the necessity to carry everywhere a handbag 'in this day and age'. On this matter, Lady Maitland had been insistent, her indoctrination had been successful and as a result, Prunella's permanent appendage was 'The Kelly' by Hermès, a bag of particularly generous proportions. Never had she felt so grateful for Mummy's advice. Two jars of Compound X were deftly inserted into her Kelly, which accommodated both without the slightest hint of a bulge. Prunella casually strolled back towards the rat cages and was giving every appearance of surveying them intently when Simpson returned from his phone call. She had more questions.

'How much did you say you give them of Compound X?'

'How much does your average rat weigh?'

'Is that fat or thin?'

'And you just mix it in with their food?'

'Grams and kilograms don't mean much to me'. But she thought it wise to stop this line of questioning.

'Let's get back to the others. This is fascinating work, Dr Simpson. We're so proud'.

Prunella and Franklin IV went to lunch with President Winthrop. The result was that the Faculty of the Todd Institute for Moleculetics were directed to focus their research on Compound X and its ilk. Only Tom Carroll refused, claiming that the work was misguided, irresponsible and that their concussions were erroneous. Simpson threatened to suspend Carroll; Carroll threatened to denounce the work publicly. Simpson decided that an accommodation could be reached that would allow Carroll to continue with his own research, while refraining from commenting publicly on the topic of molecular agents that modify the appetite of rats. He did not, however, refrain from commenting freely in the privacy of the Group Meeting. The phrases 'unsound experiment', 'sloppy methods' and 'deeply flawed conclusions' had sprung easily and frequently to Tom's lips in recent months. These had no effect on the direction of the de Lucca project; nor did it stem the seemingly endless flow of 'short communications' that appeared in *Transactions*, but this combination of private noisiness and public silence did have the important effect of allowing Tom to proceed with his own research with a minimum of interference from

Simpson.

In any case, Joe Simpson had his hands full managing the Prunella-related consequences of the de Lucca work. He was now hounded by President Winthrop for progress reports, and dragged into meetings with the medical faculty of the Bear School of Cosmetic Surgery. It was worth it, though. The flow of funds for equipment and other resources increased. Only one request did the President refuse. He declined to allow Simpson to test his compounds on beagles, which he had wished to import by the truckload. Franklin J Todd IV had expressly forbidden the use of dogs in research.

Prunella Todd, the authoress of Joe Simpson's latest burst of good fortune, was born and raised in Scotland, but had lived on the West Coast for 30 years. Like many Californian women of her age and class, she regarded Nancy Reagan as a role model, and was constantly striving to emulate the First Lady in the way she managed both her husband and her weight. In the former, Prunella had been a resounding success. Franklin J IV was a model of uxoriousness and could refuse his 'Prune' nothing. Weight management was more of a challenge, however, as in this department Prunella was less of a 'Nancy' than a 'Barbara', the figure so amply embodied in Mrs Bush Senior. Many weapons, both physical and chemical, had been brought to bear in the war to transform Prunella's Barbara into a Nancy: with only limited success. Mrs Todd loved her food, and weight control was a constant struggle. Her delight on learning of the de Lucca work was both genuine and sustained.

'That was a most inspiring visit, Frank! We should make sure they get the resources they need', she declared as soon as they settled into their private jet for the flight back to Santa Barbara.

'What do you mean, Prune?' 'I think that we have done quite enough development in that windsurfing facility. It's meant to be an Institute, not a theme park'.

'I agree, but that's not what I'm talking about. That fellow Simpson has at last done something useful with all those millions we throw at him and his lab'.

'Oh, what's that?'

'I do wish you'd pay attention. That stuff he's invented that makes rats lose weight. Just think of the possibilities. No more diets. No more liposuction'.

'Oh that', said Frank. Rats aren't people. Do you know how long it takes to bring a new drug to market. Ten years, at least: they have to do a lot of tests. You'll be getting on for 70! Anyway, I think you're fine as you are. I like you cuddly', he said sweetly.

'Well, I don't! And unless I do something drastic I'm going to grow too cuddly, even for you'.

'Nonsense! You'll never be too cuddly for me'. And he leant over and gave her a cuddle, just to prove it.

'Oh, Frank, you're the sweetest man in the world and I love you, but I want to lose weight before I hit 70. This Compound X that Simpson's fellows have come up with could do the trick. We must make sure this work is properly funded'.

'Whatever you want, my precious'. Franklin Todd IV knew that there was no point in arguing with his wife when she was in this kind of mood. He could sense another costly fad coming on, but he was not alarmed. He would speak to Mitch, Prunella's personal maid, when they got home. Mitch would know how to handle this, she always did. This perked him up, and he decided to assert himself on one thing. 'Simpson mentioned that they were planning to test this stuff on dogs. No way does any scientist experiment on any dog in any institute funded by the Todds. I already told Winthrop '.

Just as Frank knew that Prunella's weight reduction efforts were not up for discussion, so Prunella knew that on the subject of canine welfare, her husband was equally intransigent. He had taken over from his late, adored mother as the principal benefactor of the Santa Barbara Dog Home, and the only competition Prunella had for his affection was from his pet labrador, Missy. She couldn't resist a dig.

'Missy could do with losing weigh too'.

'I like my dog like I like my woman', replied Frank, bringing a reluctant smile to his wife's face.

When Mr and Mrs Franklin J Todd IV arrived at Elysium, their Montecito home, the usual welcoming party of Desmond, the butler, Missy, the labrador and Mitch, the maid was there to greet them. The Todds went their separate ways. Franklin, having exchanged a meaningful look with Mitch, made for his library, followed by Desmond, who relayed various messages *en route*, and Missy, who despite her considerable *embonpoint*, skipped merrily around her master's feet, delighted to be re-united with him. Prunella headed upstairs, Mitch bustling behind her. As usual, she was sharing with her maid the events of her day. Already alerted by a meaningful look from Mr Todd, Mitch's highly tuned antennae could sense a new enthusiasm coming on and her rapid mental processes were already beginning to lay the foundations of a plan to limit the damage. The first foundation stone was, as always, 'information', and Mitch would obtain as much of this as she could between now and the cocktail hour.

Mrs George Mitchell had been with Prunella Todd since the summer of 1943. For almost five decades, 'Miss Prunella' had been her special study and life's work. The late Mr George Mitchell had been batman to Prunella's father, Brigadier Sir Hector Maitland, during World War Two. George had, regrettably, perished in the North African campaign, leaving a widow, six months' pregnant . Both Mr and Mrs Mitchell's families had worked for generations on the Maitland estate in Perthshire and Prunella's mother, a determined woman, 'took in' the widow, assisted at the birth of her baby boy, and quickly made her an indispensable member of the household. Lady Maitland would not hear of Mrs Mitchell leaving, for reasons that were not entirely selfless. Despite her many formidable talents, domestic chores were a mystery to the chatelaine of Nethie Castle. The kitchen and the laundry were uncharted territory to her, and she would sooner traverse the Hindu Kush than the green baize door that separated the servants' part of the house from their masters. Until 1939, this had not presented a problem, but since the outbreak of hostilities, all of their able-bodied servants had been actively engaged in the War Effort. As a result, Her Ladyship had been forced into many close encounters with cooking utensils and other unfamiliar forms of domestic equipment. The results were not encouraging, and the smell of burnt food had become all-pervasive. Fortunately, for most of this time, the regular castle population was somewhat depleted. Her husband, Sir Hector was in the Army, her four boys were away at boarding school, and the war had brought a halt to entertaining on the grand scale. Lady Maitland's domestic talents were, as a result, largely practised on herself and her daughter Prunella, and confined to the one wing of the castle left to them by the civil servants who had taken over the bulk of the ancestral pile 'for the duration'.

'Mitch' as she was known, having inherited her husband's army nickname, quickly became indispensable to Her

Ladyship. She had many talents, both indoor and outdoor. She could milk a goat with as much ease as she could bake a soufflé. But one talent set Mitch apart from all others, and was the secret of her special hold over Lady Maitland. She could handle 'Miss Prunella'. Not that Miss Prunella was a difficult child. It was just that she had four brothers and a mother who worshipped boys, but was uninterested in girls. Freed of maternal responsibilities, Lady Maitland could devote herself entirely to the War Effort and to the Estate, her two absorbing interests. During those years an unbreakable bond between Mitch and Prunella was formed. Behaviour that Lady Maitland found 'difficult' – a lack of interest in ponies, hunting and fishing, and an unhealthy interest in cookery and reading – Mitch found charming. The two of them would wile away the hours in the kitchen, baking and telling each other stories. But their strongest common bond was Mitch's son 'Wee Geordie', whom Prunella adored. While the other little girls had only dolls to play with, Prunella had the real thing, and the proudest day in her young life was when Mitch allowed her, at the age of 5½, to push Geordie around the garden in his pram, by herself!

As Prunella grew up, Mitch remained the major influence in her life. The end of the War and the departure of the civil service had not brought an end to lean times at Nethie Castle. From Lady Maitland's point of view, the twin curses of the servant shortage and a Socialist government were making life very difficult indeed, and had it not been for Mitch's presence as housekeeper, cook, nanny, lady's maid and general factotum, the Maitlands could not have continued living in the castle . The cry 'Mitch!' was by far the most common utterance among the Maitland family in those days. But Mitch never complained. Lady Maitland had given her and her boy a home, which was a debt that she felt she could never fully repay. By the 1950s things became easier at Nethie. Sir Hector's war memoirs, *'Behind Enemy Lines'* had sold well and were made into a film starring Ray Milland. This had greatly improved the family finances. Long overdue repairs were done to the castle, the gardens were restored to their former glory and a new generation of servants recruited. Mitch, who had held unchallenged domestic power at Nethie for over a decade, and was by nature an absolute ruler, did not take kindly to incursions on her authority. Territorial disputes with the new butler and cook became so wearing for all that she was forced her to retreat to her last redoubt, Miss Prunella, where her writ still ran unchallenged. The daughter of the house selected no dress, took no trip and made no change in her life whatever, without first consulting Mitch. Which is not to say that she invariably did as Mitch said; she did not. But the consequences of such departures from orthodoxy were usually so unpleasant that deviations were rare.

In no aspect of Prunella Maitland's life was Mitch's influence stronger than in her choice of 'young men'. Her debutante year had been a great success. Her vivacity and beauty had made her much proposed to by the eligible bachelors of her day, but she would find that she did not really love them, usually after a discussion with Mitch, and the result would be a firm 'No'. Her mother's frustration grew as each year passed and her daughter remained unmarried. When Prunella reached the grand old age of 27, having declined at least three titled suitors, Lady Maitland began to panic, and began to suspect Mitch's motives. They had their first serious row in two decades.

'Mitch, you don't want her to marry, do you? You'd lose your influence over her', was Lady Maitland's opening salvo.

'I only want her to be happy, Your Ladyship', countered Mitch.

'How will being an old maid make her happy?'

'Miss Prunella will get married, but not to one of those chinless wonders you have in mind. She's too good to waste on them'.

'They're some of the best families in the land. Prunella could have had a lovely life with any of them'.

'Most of them are cretins. Too inbred by half!'

'Remember your place. That's no way to talk about your betters'.

'I'm not talking about my betters. I'm talking about some half-educated boys who would bore your daughter to tears. They're not what she wants!'

'Don't tell me what my daughter wants!'

'If I don't, who will? You hardly know that girl. You've never spoken to her for more than ten minutes at a time. Lady Maitland, I admire you more than any other woman I know. I've seen how you kept this place going during the War, I've seen you capture a German parachutist single-handed, and I've seen how you've helped every family on this estate when they've been in trouble. I'll never forget how you took me in when I was widowed. I've also seen how happy you and Sir Hector are. Don't tell me you married him for his little baronet title. You told me yourself that you turned down two Earls and a Marquis'.

'Yes, but that's me'.

'And you'd find that Miss Prunella is her mother's daughter, if you would take the trouble to get to know her'.

At this point Lady Maitland ran up the white flag.

'I suppose you do have a point, Mitch. That will be all'.

'Very well, Your Ladyship'.

'By the way. You know I'd have never kept that parachutist pinned down if you hadn't arrived on the scene with that frying pan and banged him on the head'.

'Best use a frying pan was ever put to, Your Ladyship', chuckled Mitch, and the storm passed, to the relief of both protagonists.

And Prunella's love life proceeded unmolested by maternal interference. Allowed to prospect in her own way, with only occasional strategic guidance from Mitch, she eventually struck gold, literally, in the form of Franklin J Todd IV, heir to the Todd mining, railroad and property fortune.

The place was St Moritz, where Prunella was skiing with friends, and the incident was a collision on the piste. Franklin, skiing much too fast, as usual, had collided with a tree, knocking himself out. The day was foggy and visibility poor, and it was only by good fortune that Prunella came upon him as she picked her way slowly down the slope. Finding an Adonis laid out cold in the snow, she sprang immediately into girl-guide mode, checked his pulse and put her ski jacket over him to keep him warm. She slapped his face and he began to moan. 'Good, he's conscious', she thought. She began to scream for help with her considerable lung power. He opened his eyes, took in her dark hair, fair skin and blue eyes, and was smitten.

They were married six months later. Lady Maitland was delighted that Prunella had made such a spectacular catch. Always practical, she felt that the lack of a title was more than compensated by the plenitude of dollars. Mitch was delighted that she would now escape from Nethie Castle and re-establish an independent fiefdom with Miss Prunella in California. There was, of course, no question that she was going. She would never leave Miss Prunella, she liked Mr Franklin, and was looking forward to the adventure. Wee Geordie had finished his studies and was now a young lawyer living in Edinburgh, so she had little to keep her at home.

Mitch's new domain in California was much to her liking. The Todd household was a happy one. Above stairs, four little Todds appeared in quick succession; two boys, two girls, two dark, two blonde. All delightful, and devoted to Mitch. Below stairs, no domestic was appointed without Mitch's express approval, and over the years her rule became more benign. Desmond, the butler, was a model of subservience and she almost admired the way that he would unquestioningly follow his master's instructions to the letter, no matter how absurd they were. Likewise, the cooks, the chauffeurs, the housemaids and the gardeners. Mitch had housetrained them all. The only part of the household she continued to regard as her own personal property was Miss Prunella.

The death of the eldest Todd child, Franklin V, on safari in Africa, had been a great shock to Prunella, who spent many hours crying on Mitch's shoulder. For weeks she was inconsolable, but, eventually, she came round to Mitch's point of view, that 'You can't wrap them up in cotton wool, and he died doing something he loved'. His father, coming as he did from the pioneer stock that had suffered more than its share of untimely demises in the course of Winning the West, took it better and persuaded Prunella that 'Frankie's' life, though short, had been full, and they should turn their thoughts to how to best commemorate him'. This whetted Prunella's appetite for 'Projects', one she had inherited from her mother, and she set to thinking about how some of the Todd billions could be used to create a worthy memorial to her son. The result had been most satisfactory, and the glimmering pyramid that housed the Todd Institute for Windsurfing as well as the Todd Institute for Moleculetics warmed her heart whether she saw it by day, when it reflected the bright Californian sun, or by night, when the lights of the ever-industrious laboratories shone over the surrounding reflecting pools, into the small hours. The beacon that was her Frankie would shine 24 hours a day.

Now, as she sat at her dressing table, Prunella thought how nice it was that the research going on in that gleaming pyramid would bring further lustre to her son's memory. It would also impress her friends, who were in the habit of changing the subject when she mentioned The Todd.

'They're just jealous', she thought, as she brushed her hair.

'Nobody goes to that art gallery downtown that Muffy Lincoln funded'.

'And the goings on at Shelly Borman's Centre for Substance Abuse hardly make polite cocktail party conversation'.

'I suppose that Betsy Lee's Foundation does a lot of good work in the Third World, but it's not something you can show off on prime time TV'.

Yes, her three regular bridge partners were all jealous of the success of the Todd Institute for Windsurfing. Barely a week passed without it being mentioned in some branch of the media. It had won so many awards that she now resolutely declined invitations to attend dinners to receive them. She made sure, however, that these invitations and the attendant press clippings were prominently displayed when she was At Home to her friends. If they were jealous now, just wait till she told them about Compound X. All three were ardent warriors in the battle of the bulge and between them had tried every diet known to man. How convenient for Prunella that dieting was an evergreen subject at their regular, weekly bridge afternoons. It would not be difficult to steer the conversation round to the latest research breakthrough at the Todd. She doubted whether they would want to change the subject quite so quickly this time.

'Will there be anything else, Miss?'

'Mitch? I didn't know you were still here'.

Mitch was indeed still here, and had been watching Prunella closely since her return. She had already established that whatever Prunella was up to had something to do with the visit to the Todd Institute and had deduced that there was something of importance in her handbag. This was usually tossed aside as soon as the mistress came in, but today she held onto it as she moved around the room and it now sat, unopened, in front of her on the dressing table. Mitch had further concluded that Missy was involved, as Prunella had been expressing an unusual degree of concern at her weight and wanted Mitch to find out, to the nearest ounce, what it now was.

'Look, Mitch, about Missy. It's not good for her to be so fat'.

'It's not good for her to eat all that junk food that Mr Franklin gives her'.

'Well, he won't listen to me about it. Do you think he might listen to you?'

Mitch knew he would. Franklin Todd IV always listened to Mitch, but she preferred to use her influence 'sparingly' and was reluctant to expend it on matters canine.

'Why me? I know nothing about dog nutrition. Or so you tell me every time I point out that titbits aren't good for those pekes of yours'.

'Mitch, my pekes aren't fat. Now, listen'. Prunella

held forth about the wonders of Compound X, explained to Mitch that it was a new animal food supplement that helped them lose weight and produced one of the two jars from her handbag.

'Look!'

Mitch looked and did not seem to be too impressed.

'What do you want me to do with that?'

Prunella Todd told her.

Chapter 6: Sir Henry Has His Say

Tom Carroll's paper was very well received at the International Congress on Moleculetics in Tallinn. Sir Henry Wiseman was in the audience, sitting, as was his habit, in the front row, directly in Tom's line of sight. This honour from the 'father of the field' was not necessarily welcomed by the speaker. Sir Henry was known, even on the verge of his eighth decade, for his penetrating questions, his ability to spot the flaw in any argument undiminished by the years. An encyclopaedic knowledge of the scientific literature, updated daily by a prodigious reading programme between the hours of 4.30 am, when he rose, and 8.00 am, when he joined his wife for breakfast, combined with deductive powers matched only by the famous resident of 221b Baker Street, London, made him a daunting interrogator. Many a new theory had shrivelled in full public view beneath the intense glare of Sir Henry's intellect. Tom's, however, blossomed. So well did he hold up under questioning that at the end of his interrogation, Sir Henry accorded him the rare public accolade 'very original work' and suggested some new lines of investigation he might consider to further his research.

Just as much a fixture in the front row at such meetings was Sol Jacobson, Sir Henry's fellow Nobel Laureate and occasional intellectual sparring partner. The 'Sol and 'Henry' show had been a fixture at the ICOM meetings for three decades; they had known each other since they had been eager young researchers in the same laboratory at Harvard. The intense rivalry of their early years had matured into a mutual regard and, since their joint award of the Nobel Prize in the 1970s, into real friendship.

'Very good work that, Sol', murmured Henry to his neighbour after Tom sat down.

'Yes, Tom Carroll's one of our rising stars. He's publishing some very good work'.

'All in *Acta*, I trust!'

'Of course, Henry'

Although both Sol and Sir Henry had retired as Editors of *Acta Moleculetica*, they continued to take a paternal interest in the welfare of the journal they had co-founded three decades before.

'Tom Carroll would be an excellent addition to the department at Yale, Sol. You could do with some new blood there'.

'Your right, and he'd be a great catch, but he seems set on staying in California. I've asked him informally'.

'California? Where?'

'Bear'.

'Where?'

'Bear!'

'Bear? Isn't that the place that Fyfe's always banging on about? What's a nice chap like Carroll doing in a place like that?'

Sol told Sir Henry about Bear, the Todd Institute and

the vast numbers of papers now cascading from it into *Transactions*.

'I am happy to say that I never look at *Transactions* beyond the contents page, which invariably confirms my view that it has published nothing of value in two decades. All these short communications and never any experimental detail! Where's the beef?'

'There's a very interesting story behind that, Henry'. But Sol was interrupted by the session chairman, who rose to introduce the next speaker. His pleasure at welcoming Professor Leslie Fyfe was not shared by the two distinguished gentlemen in the front row. Immersed in conversation, they had omitted to make their customary exit before Fyfe's presentation. It was now too late. Both groaned, and steeled themselves to sit through the ordeal that now loomed. Fyfe was notorious, both for the dullness of his research and the verbosity of his presentations. He was equally notorious for using such occasions to plug *Transactions*, the journal of which he had for the past five years been Editor *ad interim*. Today he plumbed new depths. Fyfe chose to speak, not about his own research work, but about a series of short communications published in *Transactions* by the Simpson research group at Bear State University. It was his most shameless plug yet, and the audience quietly seethed. Only Sir Henry seemed indifferent. His thoughts were already elsewhere, analysing a scientific problem that had been perplexing him for some time. 'You will be pleased to know', asserted Fyfe, 'that the impact of *Transactions* has grown hugely in the last two years'. Sol nudged Sir Henry, who, roused from his thoughts, began to pay attention. This was not good news for Fyfe, who proceeded to show how the increase in the impact of *Transactions* could be attributed to a series of excellent 'short communications' from Bear

State University that introduced a new group of molecules that, they claimed, acted on the hunger centres in the brains of rats and suppressed appetite. Their proposed mechanisms of action were described. 'Based on what evidence?' murmured Sir Henry to himself. Fyfe further attributed the fact that *Transactions* had attracted these papers to its 'rapid publication process, twice as fast as the major competitor'. 'He publishes whatever comes in, doesn't bother with peer review', whispered Sol to Sir Henry.

'*Transactions*' policy, unlike that of many other journals, is to be inclusive'. Sir Henry replied, *sotto voce*, 'Inclusive, yes, you'll include any old rubbish', but understood well that this was a specific dig at *Acta* and its reputation for rejecting over 80% of the papers submitted to it. Sir Henry did not rise to the bait. It had been his firm policy over the past three decades to ignore the outpourings of Les Fyfe and his apostles. He did not read their papers and studiously avoided their conference presentations. Sir Henry's rapid exit from a room whenever Fyfe rose to speak had become one of the regular rituals at conferences in the field. This had preserved his peace of mind. His reserves of equanimity were, however, evaporating rapidly, as Fyfe, enjoying his captive audience, and emboldened by Sir Henry's presence, continued to sing the praises of *Transactions*, and made increasingly obvious digs at *Acta*. Eventually he went too far.

'*Transactions*, unlike other journals, is not elitist. We believe in facilitating communication among scientists, not obstructing it'. At this, Sir Henry broke the habit of decades and stood up: he had a question for Fyfe. All eyes turned towards him. The session chairman, who was familiar with Sir Henry's settled position on the matter of Fyfe, was temporarily struck dumb. Other hands, raised inquiringly, fell. People intent on making their escape from the room turned back. Silence fell. Fyfe's chins wobbled.

'Professor Fyfe', began the great man, 'You claim that a series of short communications, published in *Transactions in Moleculetics,* of which you are Editor *ad interim*, provide strong evidence for a new class of molecules that have the effect of suppressing appetite and lead to weight loss in rats?'

'Yes, strong evidence'.

'What evidence? As I understand it, no experimental details have been published that confirm how these materials act on the brain centres, which makes the mechanisms proposed purely speculative. When will we see the evidence?'

Fyfe's chins wobbled a bit more. 'Full papers with experimental details will be published in due course'.

'In other words, you have no experimental evidence'.

The wobble of Fyfe's chins increased further. He began to bluster.

'Not true. The molecules in question are well characterised, and similar mechanisms are described in the literature'. This was a major tactical error. If Sir Henry knew anything, he knew The Literature.

'Similar mechanisms do indeed exist, but only in extreme conditions of high temperature and pressure that are impossible in any brain. Are you familiar with the work of Von Straub and Uberoff? Their results would indicate quite another process: one that has little to do with hunger centres'.

'Of course I know their work, but that was a long time ago. Things have moved on'. The chin wobble grew. Fyfe knew the names of these two distinguished German scientists, but not the details of their research. He did know better than to challenge Sir Henry on his knowledge of the scientific literature and decided to change tack. The force of his bluster grew.

'There has been a very broad acceptance of the results reported in *Transactions*. The papers are widely read and very highly cited by other scientists'.

Not for nothing had Sir Henry served in His Majesty's ships during the World War II. One who had braved the North Atlantic in winter was not going to be blown off course by a minor windbag like Fyfe.

'Highly cited? Where and by whom?'

'That's an impossible question to answer'.

'Not so. It is easy to identify the authors who cite this work and the journals in which they publish. As you speak with such authority on citations, I assume you must be familiar with citation data and where you can find it'.

'Of course I am', replied Fyfe, 'but I do not have the data to hand'.

'Then I shall ask you for something that you must have to hand as Editor of *Transactions*, albeit *ad interim*. You are, I take it, familiar with the concept of peer review. How many people review these short communications before they are published?

'They have to be published quickly, that's the whole point'.

'That may be the point, but it does not answer my question. How many reviewers?'

'That's confidential information', replied Fyfe, whose bluster was now fading fast.

'Rubbish. *Acta Moleculetica* uses three reviewers, apart from the Editor. How many reviewers do you use?'

'One', asserted Fyfe, 'this process has to be fast'.

'Apart from the Editor?' enquired Sir Henry.

'Usually', replied Fyfe.

'But not always?'

'No, this process has to be fast'.

'Publish in haste, repent at leisure', quipped Sir Henry. This prompted the first ripple of laughter in the room, which had, until now, held its breath as the tension of this exchange increased. The audience had grown considerably. An interrogation by Sir Henry Wiseman was a highlight of any conference and it had been some years since he had fully unsheathed his rapier. Veterans of ICOM meetings, many bearing scars from previous Wiseman interrogations, recognised that he had lost none of his sharpness. Fyfe, who had until this moment thought himself immune to interference from 'that man', could see this too, but his manner had by now become that of a mesmerised rabbit, rooted to the spot. He looked in panic towards Vera. Alone among the throng, she was stony-faced.

'Let me get this clear', mused Sir Henry 'for many of these papers you are the sole reviewer?' The audience gasped. It was widely suspected that the peer review process, so central to the integrity of the scientific literature, was rather loosely observed by the Editor *ad interim* of the *Transactions in Moleculetics*, but the bald, public confirmation of this suspicion was shocking.

'Yes, to speed up the process...'. Fyfe's voice trailed off.

'You've made that point already. I do not follow your work closely' (laughter from audience), 'so you will appreciate that I was unaware you had become an expert in neurochemistry. Are you now working in this field?'

'No, but chemistry is chemistry'.

'And medicine is medicine, but I think if I had a heart condition I'd want the opinion of the appropriate specialist

– a cardiologist, not a gynaecologist.. (Loud and hearty laughter.) What about the other reviewers?

'What about them?'

'Where do you find them?'

'We have an excellent editorial board'.

'Any neurochemists?'

'Well, not exactly…'.

'But chemistry IS chemistry, I suppose?' Much further appreciative laughter. Sir Henry was on vintage form, and now turned his firepower on another weakness in Fyfe's defences.

'Although I am now retired, I still do my best to keep up with the literature. It is a lifelong habit I find hard to break. As far as I can recall, I have not seen this work widely cited in the chemistry journals that I read. I wonder, therefore, where all these citations you refer to are coming from? Is it the neuroscientists, perhaps, given the centrality of the brain in this research?'

At this point, Tom Carroll, who was sitting at the far end of the front row, punched the air and hissed 'Yesssssss'. He could see that Sir Henry was on the scent.

Fyfe had a ready answer. 'Actually, you will find many citations in *Transactions*', but realized that his was the wrong answer, as he recalled, too late, Sir Henry's great disdain for this journal, which he never read. The audience knew this very well, and a collective thrill ran through them as they moved to the edge of their seats. How would Sir Henry reply? He said nothing.

Then Fyfe elaborated, hastily and fatally, on his earlier remark. 'And there are a lot of citations in the nutrition/obesity journals'.

'Not a body of literature with which I am familiar, but I think it disturbing that these new findings should make their way, with little or no intellectual challenge, from basic chemistry to nutrition. Quite a journey'.

'Well, they've been tested on rats, and they lose their appetite'.

'That's the nearest to a scientific observation you have made all afternoon', was Sir Henry's final observation as he sat down. There were no further questions; nobody dared. The chairman brought the session to an end.

This had been an entertaining exchange for the audience, who left the room in a high state of excitement and looked forward to dining out on the story for many years to come. But it had been a dangerous one for Les Fyfe, and he knew it. His ill-advised provocation had piqued the curiosity of Sir Henry Wiseman. This had claimed its first casualty, Fyfe's own credibility as an Editor, when he had been provoked into revealing the slackness of *Transactions'* quality control processes. There would be more casualties as Sir Henry's inquiries progressed, as they undoubtedly would. His inquiring mind had been stimulated. Fiona Hamilton had been nagging Fyfe about the need to get full research papers out of the Bear people. He had refused, of course. No way was he going to publish the full experimental details on how Compound X and its analogues were made. At least not until the patents were secured by Pennine Nutritional Supplements, the company of which he was a non-executive director. Wiseman's interest complicated things. Fyfe assumed that he'd see the importance of this stuff and suspected that he'd soon be telling his pals in the big pharmaceutical companies. He made a mental note to call Archie Ramsbottom as soon as he got home. He and that company of his would need to get the finger out. Fyfe was pondering these matters as his wife waddled over to the lectern at the end of his presentation.

'Bloody old fool', were her first soothing words of comfort.

'I know. He shouldn't be let out on his own'.

'Not him, you! Why did you have to go and rile him like that? You let him make a laughing stock of you and the journal, right in front of most of your editorial board, not to mention Fiona Hamilton. It's the chance she's been waiting for'.

'Fool? Me? Fiona Hamilton? What do you mean woman?'

'Don't you see? Look there! Wiseman and Jacobson are in a huddle with Tom Carroll. Mark my words, they've got it in for you. Unless you act quick, it'll be Bye-bye Bear, Cheerio Journal, no more international travel and we'll be stuck in Rochdale'.

'Don't talk nonsense. This work from Bear is solid; I'm sure of that', said Fyfe, without entirely convincing himself.

Fortunately, as yet, Vera knew nothing about the Pennine connection with this research. She was upset at the prospect of losing a few thousand pounds worth of editor's fees. If she thought that hundreds of thousands of pounds, in the form of his Pennine stock options, were potentially at risk, life would not be worth living. Archie had to get moving or Wiseman and his pals will strike it rich, not them.

Vera interrupted his thoughts. 'Let's go, we've got to get ready for the Board Dinner this evening'.

Fiona had lingered in the room. Always on the lookout for promising, new authors, she had been impressed by Tom Carroll, and wanted to see whether he might be interested in writing a book for SI. When he had finished talking to Sir Henry and Sol, she introduced herself.

'From SI? Publisher of *Transactions*?'

She knew that this was not a good start to the discussion, and suspecting that the usual criticisms of the journal would be forthcoming, adopted her breeziest manner and replied: 'Yes, actually, I'm responsible for the management of *Transactions*'.

Tom Carroll launched into the traditional rant, which Fiona could have predicted word-for-word. In fact, she agreed with much of what he said. The journal IS a disgrace. Fyfe IS unfit to be an editor. The Editorial Board IS packed with his cronies. No scientist of repute WOULD dream of publishing their work in the journal. She smiled, nodded and once Tom's rant was spent, said 'I quite agree, and I'd like your advice on how to fix it. How about a drink?'

Taken aback, he replied 'Oh, all right', and they made for the bar.

Sol and Sir Henry were left alone in the empty lecture room, deep in conversation.

'The mechanism they propose is impossible under these conditions'.

'Henry, the Simpson group have published 20 short communications in *Transactions* over the last three years, all proposing the same mechanism. I don't think any research chemist believes them. Nobody takes them seriously, as you know'.

'Yes, Sol, perhaps no chemist believes them, but they seem to have taken in the nutrition people. Clever ploy that. Anyway, who is this Simpson fellow?'

'I don't know too much about him, but he has been Director of the Todd Institute at Bear since it was founded a few years back. He's British, and I believe there is a Fyfe connection. The Todd's got a huge endowment'.

'Sol, there is something fishy here, and I'm going to get to the bottom of it. Fyfe and that journal are a disgrace to our science. We can't ignore him any longer. If there is, as you say, some link between him and the Todd Institute, you can be sure that it is based on his greed and their money. I must speak to Bertie'.

Sol yielded to nobody in his admiration for Alberta, Lady Wiseman. He worshipped her many virtues, but had not until now considered that a deep knowledge of moleculetics was one of them'.

'What would Bertie know about it?'

'It so happens that we sat next to a young lady on the plane who is in some way responsible for *Transactions* at Standard International. It turns out that she and Bertie went to the same girls' school in Scotland, obviously a century apart, but my wife took to her, and you know what that means'. Sol did, having been a special pet of Lady Wiseman for decades. Indeed, it was Bertie who had kept open the diplomatic channels between the two men in the earlier part of their careers, when rivalry was intense and their forceful personalities had not yet been subject to the soothing influence of a Nobel Prize.

'So Bertie will have arranged a dinner with this young lady?'

'I think that is a reasonable assumption, Sol. Let us find out when, and get ourselves invited'.

The young lady, meanwhile, was deep in conversation with Tom Carroll. They both had much to ask of each other, although neither would admit to it. Discussing a possible book was merely the pretext for a longer chat. She really wanted to know what went on at the Todd. He really wanted to know what actually went on at the journal. They were both concerned about the growing trend towards the sensational in scientific research. The recent debacle over the alleged discovery of a 'cold fusion' process, when it had been claimed that this process would deliver vast quantities of energy at no cost to the environment, was fresh in their minds. The embarrassment of the scientists, the universities and the publisher involved in that fiasco had been very public. Fiona was unaware of Klomp's plans to launch SI into this already discredited field.

'And it could happen again', said Tom, ominously. 'Some colleges are so desperate to attract money, they'll jump on any bandwagon. They are no longer run by academics. The accountants have taken over. The President of Bear is an accountant'.

'But I thought that Bear was never that academic to begin with'.

'Ouch! Hey, but look at the number of Ursuans who have won the Masters, Wimbledon and America's Cup. The previous President won an Olympic Gold Medal for skiing. Bear aims to develop the body as well as the mind'.

'I know. *The Daily Telegraph* had an paper last week about some British students who have just won athletics scholarships there'.

'Which sport?'

'Windsurfing'.

'Of course. The Todd Institute for Windsurfing is the best facility in the world. When it's made an Olympic sport, Ursuans will sweep the Board'.

They both laughed. 'Another drink, Fiona?'

'I should be paying. I'm on expenses, and you are just a poor academic'.

'There is no such thing as a poor academic at Bear, at least financially speaking. Our travel budgets have always been generous.

Fiona changed the subject'. Who are these Todds. Don't you work at a Todd Institute?' Asked Fiona, her mind turning more forensic. This was her entrée to further probing of the Todd/Simpson/*Transactions*/Fyfe axis that had been intriguing her for some time.

Tom told the story of the Todds, from their beginnings in the goldmines, via the railroad boom to the Institute for Moleculetics.

'The *Telegraph* says that the windsurfing institute is wonderful. Is it true that there is an indoor lagoon?

'Yes, with computer-controlled wind and waves that can replicate almost any conditions. It's great'. Tom betrayed his enthusiasm for the sport.

'So you're a windsurfer yourself, then?'

'Sure, why do you think I stay at Bear? Where else could I windsurf and do research in the same building?'

'Tell me about your research'.

He told her.

'You certainly impressed Sir Henry – not an easy thing to do'.

'He asked some real penetrating questions, and wants to discuss my research with me while we are both here. I'm scheduled to have dinner with him tomorrow. I'm excited. He's so insightful, and there's nobody at Bear I can discuss my work with'.

'Why not, I thought the Todd was full of researchers'.

'If you mean Joe Simpson's group, forget it!'

'Oh, why so dismissive? They publish lots of papers in *Transactions*. They seem very productive', opined Fiona in her sweetest voice.

Tom felt himself being blown into dangerous waters. Simpson was, after all US Editor of *Transactions*; Fiona would know him, so Tom had to be careful what he said. The terms of his truce with Simpson meant that he had to keep his opinion of that group, and the work they were doing to himself. He quickly tacked out of trouble.

'They work in a very different field than me. Different molecules, different mechanisms, different techniques, different instruments'.

'OK, I get the point. They're different!' But Fiona sensed that there were other reasons for Tom wishing to avoid discussing Joe Simpson.

Tom tacked again, perhaps too hastily. 'How are things on *Transactions*? It's not a journal I have ever published in. Les Fyfe seemed very pleased with the way it is going. How does he do it?'

Fiona flared up at the mention of that man's name. 'Fyfe merely does as he is told!' she snapped.

'Told by you?'

'No, by my boss'. Fiona did not want the conversation to turn to the editorial management of *Transactions*, as this would reveal (a) how little influence over it she, the nominal manager, had; and (b) the financial objectives that really drove the process.

'Actually, the journal is doing so well because of all these high-impact papers being published by the Simpson group'.

This threw Tom. 'I know that already, I was at Fyfe's presentation. And that wasn't what I meant. Is it true about no referees?'

'Look, I can't talk about confidential journal matters', snapped Fiona.

They lapsed into silence, each afraid to raise the topic that they most wanted to talk about, as it seemed to lead inexorably to the subject they least wanted to talk about. Yet, neither wanted to leave, and they searched for a subject that might lead them away from the tricky conversational shoals in which they were now wallowing.

'Is this your first trip to Tallinn?' sprang simultaneously to their lips, causing much merriment. A more relaxed discussion ensued.

On his return the Hotel Splendide International, Sir Henry found his wife wrestling with the plumbing in an attempt to take a pre-dinner bath. Her efforts had resulted in the merest trickle of a rather repulsive brown fluid from the taps. She was not in the best of moods.

'At least in Stalin's day there was enough water to fill the bath!'

'For the few, Bertie, for the few', her husband reminded her.

'But we're meant to be The Few, Henry. Otherwise what are we doing in the best suite in the top hotel in Tallinn?'

'That's democracy, Bertie. Now listen, I've invited a young man to dine tomorrow evening'.

'And I've invited a young lady. Anything rather than an 'Evening of Estonian Folk Dancing. I remember Stalin's people urging Winston to…'.

'Never mind that. What do we do about dinner tomorrow?'

'Henry, we can either combine or we can swap. A swap would certainly be more exciting. Me out on the town with a young man, and you *vice versa*. Imagine the gossip. But my arthritic joints don't feel up to it, so I suppose we should combine'.

'Does your young lady speak English?'

'Sort of, she's from Scotland. Does your young man?'

'Sort of, he's from California'.

'Then we should all get by without an interpreter'.

'What shall we eat?'

'How about Estonian?'

'What does that mean?'

'Really, Henry, I do wish you'd pay attention sometimes, I explained to you all about Estonian cuisine on the flight over. I remember in the summer of 1943, Anthony Eden and I...'.

'I don't mind what we eat so long as the restaurant is well ventilated. If this plumbing is typical, I don't want to be confined to any enclosed spaces for the next week'.

Chapter 7: The Non-Executive Director

'Look here Archie, if you want to make a killing on this, you've got to move fast. I've just got back from a conference in Estonia, and the word is getting out about the weight-loss effects of these compounds'.

'Les, what's your hurry, we've registered the patents, haven't we? Nobody can touch this stuff without our permission. Any road, it takes time to convert the plant. It's been mothballed since we stopped production of 'Weighmore' for pigs. You can't convert it overnight to producing tons of 'Thin-Kwik' for ladies, you know. Bloody 'ell, Les, this is a big investment for us. We've never done anything for humans before, and the regulations are a bit tighter than for our four-legged friends. We've got to be inspected for hygiene and the like'.

'I know all that. But I'm telling you that Wiseman is onto this. He was asking me a lot of awkward questions in Tallinn, and he suspects there is something hot here. I bet he's already been on the phone to his pals in the pharmaceutical industry. He's a pompous old git, but he knows his science and he's suspiciously interested in this stuff'.

Archie Ramsbottom, Chairman and Chief Executive of Pennine Nutritional Supplements, sighed. He had known Les Fyfe for over twenty years. They were long-standing golf partners, and had spent many a bibulous evening in the 19th hole. At some point in these evenings, usually after his third double Scotch, Les would turn to the subject of Sir Henry Wiseman, elaborate on his many character defects, and bemoan the blight that this elderly Oxonian had put on his career. Archie was all too familiar with Sir Henry Wiseman.

'Les, have you been drinking? It's not even ten o'clock yet'.

'No, I've not been drinking. I'm telling you that Wiseman is onto this, and if we don't speed up the launch, his pharmaceutical pals will beat us to it'.

'Les, calm down! We have the patents!'

'Yes, but I've told you before, once they get to the bottom of this molecular process, they'll be able to synthesise all sorts of compounds that could be as effective, and maybe even more so. These big companies have got hundreds of scientists they can switch to a project like this overnight'.

'Look, you've told me that already. You've also told me how long it takes for a drugs company to bring a new product to market. They've got to jump through regulatory hoops and all sorts. We don't, as you well know. So let's take our time and do this proper-like. That way, we'll both make more money'.

'But what about the other nutritional supplements companies?'

'Les, you know that the human nutritional supplement companies are too pernickety to move quickly. They're terrified of new government regulations after that 'Hair-issimo' debacle. All those bald men who suddenly grew breasts as well as hair! No, they've lost their nerve for the moment. And the animal nutrition companies are all concerned with weight gain, not weight loss. We've got the perfect cover: No competitor will have any idea what we're really up to until it's too late. We've discussed all this before. Don't let Wiseman spook you . Stay calm and think of your Pennine share options!'

These words did have a soothing influence on Fyfe. On his appointment as a non-executive director of Pennine two years ago, he had been awarded the option to purchase 40,000 shares at ninety pence, when the market price was £1.20. Since then, the price had fallen slightly, but he had since been awarded options on a further 10,000 shares. His first batch of options matured in just under a year: the date was circled very boldly in his diary. This day also featured prominently in his dreams, as he pictured how rich he would be. Had this not happened to the Wisemans of this world, whose share options in pharmaceutical companies had made them rich? If they wanted to throw that money away on scholarships for Africa, more fool them. Les had set his heart on a Winnebago motor home. No more touring the Lake District in a poky caravan. His Winnebago would be the envy of every campsite. Even better, it would keep Vera happy.

Fyfe, comforted by these thoughts, replied 'I suppose you're right, but I know Wiseman is up to something'.

'And let me tell you what we're up to, Les,' bellowed Ramsbottom. 'Our process men are well advanced with the design of the plant, and they assure me that it will be operational within the six-months. Now, I understand your boffins in California are making good progress with the scale-up experiments, and that they don't see any problems there'.

'Yes, that's true'

'Now, of course, being the businessman round here, I've been turning my attention to the marketing, and we've done two things'.

Fyfe tried to interrupt.

'Now shut up and listen! First, we've registered 'Thin-Kwik' as a trademark, so that's all sorted. And second, I've hired a new PR executive to take charge of this product. She starts working with us next month, and has all kinds of new ideas. She's from Down South, name of Tiffany Beale'.

Les was impressed. From Down South? Once Archie threw his considerable weight behind a project, things began to happen.

'That's good news. What's her qualifications?' Les knew Archie's predilections and suspected that her main qualifications would be an ample bust and blonde hair.

'Don't you be bothering yourself about that. You take care of the science and I'll take care of the business. That's the idea, remember'.

Fyfe found the steamroller that was Archie Ramsbottom hard to handle when it was running at full speed, as he was by now. The only safe option was to stand clear.

'That's grand, Archie. I'll leave that to you'.

'Now then, what about golf on Saturday? The usual foursome? We need to spend some time after the game with Terry and Ken to chew over this trademark nonsense and look at the budgets'.

'Lunch at the club, then, I take it. I'll tell Vera she'll have to go to her mother's on her own. What an escape!'

'The Club? No, we'll need to eat somewhere more private than that. I don't want the whole of Rochdale knowing our business. How about my place? Mavis can rustle up something for us'.

LET THEM EAT CAKE

'And how much does she know about Thin-Kwik, or about Tiffany Beale come to that?'

Archie got Les's point immediately. 'Maybe you're right. Luncheon at my house might not be the best way to keep this hush-hush. Mavis would be telling all her friends at the bowling club'.

Les saw his opportunity. 'What about the Versailles. They've got a private dining room. I know the Maitre d'. He'd take care of us'.

'I bet he would'. Thought Archie, who had a strong foreboding that he would be paying for lunch at one of Lancashire's most expensive eateries. He had little alternative, however. His domestic tranquillity would be much better preserved if Mavis knew nothing about Tiffany Beale. 'I suppose so' he replied, without enthusiasm.

Fyfe's heart leapt for joy. There were few pleasures in his life as intense as that of eating and drinking at somebody else's expense without his wife. He made a mental note to check with Gaston that there would be lobster on the menu at Versailles come Saturday.

'See you Saturday, then. I take it Pennine will be paying'.

'Of course, Les, don't we always. See you Saturday. Make sure your short game's a bit better than last time'.

Les Fyfe and Archie Ramsbottom had what might be termed a 'symbiotic relationship'. Archie picked Les's brains and Les picked Archie's pockets. They had met at the Moss Rise Golf Club shortly after the Fyfes moved to Rochdale. Fyfe had just been appointed to his first lectureship, and was pleased as punch to have, at last, a tenured position. Archie had just taken over the family firm of Ramsbottom & Son, which had been founded by his great grandfather in 1883 and was, by the 1950s, one of the country's largest suppliers of animal nutritional supplements. Alas, neither career had prospered. Fyfe, despite several attempts to escape, remained at the University of Rochdale, and his academic career had peaked with a Professorship there. He had been the only applicant. Ramsbottom & Son, since Archie took over the driving seat, had been afflicted by his facile enthusiasm for every passing business fashion: diversification, globalisation, downsizing, outsourcing, and rebranding had all taken their toll. All costly, but had led nowhere. Even that seductive road to riches, 'going public', had not produced the instant wealth that Archie had promised his many Ramsbottom cousins. In fact, they had only narrowly avoided a complete fiasco.

In the summer of 1986, during a stock market boom, all had seemed set fair for shares in 'Pennine Nutritional Supplements' – the soon-to-be rebranded Ramsbottom & Son - to rise the instant they were offered on the London Stock Exchange later that year. Then the 'Weighmore' debacle struck. It was claimed, by a group of customers, that this new supplement for pigs was causing a bitter taste in the meat, making it unsellable. Sales of Weighmore plunged and a court case was brought against Pennine. Only fancy footwork from their star expert witness, one Professor Leslie Fyfe, had saved the day. He had shown, to the judge's satisfaction, if not to that of his scientific peers, that the bitter taste of which customers complained could not, on the balance of probabilities, be attributed to the pigs being fed 'Weighmore'. The results, he said, were inconclusive. So many other factors were present, and these would take a long time to study thoroughly. Much further investigative work was needed, and this would have to be funded by somebody. The herd of disgruntled pigmen who had brought the case, had neither the stomach or the finances for a long fight, and threw in the towel. Ramsbottom & Son had weathered the storm, thanks to the plausible scientific bombast of Professor Fyfe. But 'Weighmore' did not, and was quietly dropped from Ramsbottom's product range, as part of a 'strategic repositioning' that discarded pigs from their portfolio to focus on their core strength – poultry. The new company mission statement: 'We feed them to serve you' could be more accurately described as 'Two wings good, four legs bad'.

The public offering of shares in Pennine Nutritional Supplements was delayed until October 1987, to give the newly created marketing department time to erase the stain of Weighmore from the memories of potential investors and to build the other brands 'Laymore', 'Full Breast' and 'Peckless'. But three days after a moderately successful launch of Pennine Nutritional Supplements (formerly Ramsbottom & Sons) on the London Stock Exchange, Black Thursday struck, the market plunged, and Pennine shares had languished below their offer price ever since. Contrary to the predictions of the management consultants, the position of Pennine at the top of the pecking order in the chicken feed market, with its brand leadership and market power, had not translated into sales growth. In fact, they had still not made up for the sales lost when the strategic decision had been taken to get out of pigs.

The consequences of this decision were often at the forefront of Archie's mind, usually after a Board Meeting, when one of his relatives would bemoan the fate of the Pennine shares, or after an altercation with Mavis, his wife, when she would remind him that the new pool for their villa in Spain was still only a blueprint. He was spending far more on marketing than ever before, yet his factories were running well below capacity. Whatever they did, they could not seem to increase sales of chicken supplements. Perhaps four legs weren't so bad after all.

The one bright spot of these last three years had been the appointment of Les Fyfe as a non-executive director of Pennine. It was, Archie explained to his cousins on the Board of Directors, a just reward for saving the company, and they had to become more scientific about what they were doing. After Weighmore, nobody disagreed with this, and there was much Ramsbottom mirth when Archie informed them that Les would cost the company virtually nothing. All he wanted were very modest expenses, and share options. 'So he only benefits if we all benefit'. Les had more than earned his keep on the Board, supporting, as he did, every one of Archie's plans, and intimidating the more unruly cousins with his professorial status. He had also advised them how to make some of their production processes run more efficiently, which had allowed them to reduce staff and shore up profits.

But still the share price languished. They had laid off staff, slashed other costs and had even mothballed the Oldham plant, concentrating all their activities in Rochdale and Burnley. There was not much more retrenchment to be done. Then, at one Board Meeting, where the faces around the table were particularly glum, Les Fyfe mentioned the word 'Diversification'. A collective groan greeted this remark. Anyone who had sat on the Board of Ramsbottom & Son these past two decades - and most of them had – would recall that Archie's grandest follies had come under the heading of 'Diversification'. Cousin Fred was the first to speak. 'No, we've had enough bloody diversification in the past. Us should stick with what us knows'. 'Quite right', chimed in Cousin Annie, 'No more safari parks'. 'And no more retail outlets', added Cousin Albert. 'Us has been only one step from the poorhouse with all that nonsense'.

Archie had warned Les that there would be opposition, so he was prepared. His plan was to blind them with science. Les was armed with a Presentation. The Ramsbottoms sat in awed silence as the lights were dimmed, the blinds closed and a projector burst into life. None of them had been exposed to a full-blooded Presentation before.

'Let us give the professor a hearing, shall us?' counselled Archie. They all nodded.

'The professor presented. His slides were an orgy of overdone graphics that mesmerised a board whose only previous exposure to this level of audio-visual vulgarity had been the Blackpool Illuminations. They completely missed the point of the presentation, which was the point; but by the end, their senses overwhelmed, they had lapsed into a state of supine acquiescence.

'And so', Fyfe concluded, 'the opportunities for Pennine in the rapidly growing market of human nutritional supplements are clear. You have the skills and the infrastructure, and you've got expensive plant standing idle. The research I have described to you has been done at a leading laboratory in California and published in one of the world's leading journals'. Women the world over want to lose weight, and a dietary supplement based on this research will help them do so. The potential market is huge, and I don't need to tell you the likely effect on Pennine shares'. At this, there was a collective smacking of lips around the table. Archie knew his moment had come.

'So, my fellow directors it appears to me that the professor here has presented Pennine with a golden opportunity. He's got the know-how and we've got the plant, which is lying idle'.

'But us has had us' hands burnt with diversification

before', pointed out Cousin Fred.

'True, but this isn't really diversification. Our core business is producing nutritional supplements for birds. Aren't women 'birds' too? We know what we're about here.

'I reckon it's a good idea', declared Cousin Albert. 'Until now we've been about weight gain; now were moving into weight loss. It all amounts to the same thing. I remember trying that Full Breast stuff out on my Ada; it worked a treat, I can tell you. Humans or animals, it's all chemistry, as the professor has pointed out'.

'You never fed your Ada that stuff did you?' asked Cousin Annie.

'Why not? It rounded her off a treat. She were getting scraggy. Helped her mood too'.

'But she dropped dead of a heart attack!'

'Heart trouble runs in that family'.

'Can we please get back to the point here. We have an important decision to make', said Archie, calling the meeting to order.

'I put it to the Board. Do we wish to develop this product?'

After some questions about costs, on which reassurances were given, the Board voted unanimously in favour of Pennine developing 'Thin-Kwik'. That was three months ago, and for once, things were going according to plan.

The round of golf that Saturday proved a frustrating experience for Archie and Les. Neither played well, blamed the wind and each other as they conceded hole after hole, and went down ungraciously to defeat at the hands of Terry Sharpe (Archie's lawyer) and Ken Swindley (his accountant). When they arrived at Restaurant Versailles it took two rounds of large pre-lunch Scotches to restore the atmosphere of *bonhomie* with which the game had begun.

'That pitch to the green at the 9th were very bad luck, Archie', commiserated Terry.

Ken: 'I've never seen the greens run so fast here. Your pitch from the bunker were perfectly judged. It ran across that green like Lynford Christie. Straight into t'other bunker'.

Terry: 'I blame that new green keeper. He's trying to make this place like bloody St Andrews. Bunkers everywhere and greens you could skate on'.

Ken: 'And that double bogie you had at the 10th were most unlucky, Archie'.

Terry: 'Most unlucky'.

'Alright, alright, let's not dwell on it'. Even Archie had a limited appetite for the sycophancy of Ken and Terry. No doubt the various Ramsbottom enterprises put a fair bit of business their way, and they never ceased telling him that he was their most important client, but enough was enough. He paid them plenty in fees and did not expect them to dwell on his failures.

'You both seemed to cope well enough with the greens and the wind. Must be all that practice you get. Happen we'd all be better off if you spent less time on the golf course and more time on my affairs. Speaking of which, you know why we're all here today. Thin-Kwik. Let us order luncheon and get down to business'.

All four perused the menus. A silence descended. Archie noted that today's special was lobster, and surmised that Les would surely have that. His eagle eye also spotted 'Beluga caviar' among the list of starters, and wondered whether Les would have the gall to order that, at £18! He made his own choice of soup followed by roast beef and noisily shut the menu.

'Very fancy menu today, Gaston. Are you expecting the Queen Mother?'

'No, Sir', replied the maitre d', 'We had a special delivery of shellfish'. Gaston caught Fyfe's eye. 'We have also upgraded the wine list'.

'Oh?' muttered Archie, now regretting the tradition whereby he allowed Fyfe to select the wine on these occasions. Fyfe was studying the wine list closely.

'A great improvement. My compliments, Gaston', he declared on closing it'. What are we all having to eat?'

Terry: 'Lobster'.

Ken: 'Lobster'.

Archie: 'Beef'.

Les: 'Then a nice, refined red will be perfect. A bottle of Chateau Margaux 1982, Gaston, if you please'.

'The wine list here HAS been upgraded', chuckled Terry.

To Archie, who was no oenophile, this meant only one thing. It was going to be expensive. So be it. He would make sure he got his money's worth out of them.

'So, gentlemen, the topics for discussion today are: the patent situation, progress on the scale-up, the cost of reopening the Oldham plant, and the launch plan for Thin-Kwik. Terry, how are we doing on the patent front?'

'Couldn't be better, Archie. The patents have all been registered in the name of Pennine in the UK and the applications are proceeding smoothly in the US'.

'That sounds very straightforward. Are the people at Bear happy with that?'

'Delighted. The Dean of Research seems thrilled that somebody actually wants to exploit their scientific discoveries. Apparently it's never happened before'.

'Don't they want any royalties?'

'Far from it. He asked me if they would have to pay US anything. I don't think they know what intellectual property is, and they certainly don't have a department to deal with it. The Dean told me that his secretary would take care of everything. I faxed the assignment letters immediately – had to strike before they wised up. A day later she called me to let me know that the Dean had signed the letters and faxed them back. She was very nice. Quite an expert on fax machines'.

'They all seem to be very nice out there. No wonder you travel to Bear so often, Les. Is your friend Simpson happy with all this?'

'He'll do as he's told. He stands to make a lot out of his private arrangement with us'.

'Money talks!' bellowed Archie. 'This is wonderful. No royalties to pay! That's one cost you can take out of the budget, Ken. This looks better all the time! Happen we can maybe afford that Chateau bloody Margaux after all. Tell me, Terry, are you sure these faxed documents are legal?'

Terry: 'Absolutely, there have been test cases. Any road up, she sent the signed originals by courier and they arrived a day later. They're sitting in my safe'.

The golf game by now forgotten, it was in the highest of spirits that Archie moved onto the next topic.

'Thank God for misguided efficiency! Now, Ken, tell us how much will it cost to retool the Oldham plant for Thin-Kwik'.

'That depends', said Ken.

'It always depends with you bookkeepers. Depends on what?'

'Well, there's a new set of European Union regulations on food processing hygiene just come out, and all new plants will have to conform to them'.

'Hang on there, who's talking about a new plant? We're simply reopening an existing plant. I've dealt with these people before. They always give you a couple of years to upgrade existing facilities. By then we'll have made our millions'.

'That's true, but you are changing from animal supplements to human supplements'.

'Our hygiene standards have always been the highest. Sterility is our byword! What say you, Mr Lawyer? Will we have to upgrade?'

'That depends'.

'Not another one!'

'That depends on the samples. You have to send some samples for testing , and if they pass, you get your two-year grace period'.

'Fine, we'll make sure they pass'. Archie winked at Les.

'Now, I take it that if all we are just making a small modification instead of a wholesale upgrade of the plant, the costs will not be ruinous'.

'That is correct, Archie', and Ken proceeded to take them through the cost projections.

'Most satisfactory, most satisfactory', declared a beaming Archie.

'Luncheon is served', announced Gaston'. May I conduct you to our private dining room, the *Salle Trianon*'.

'Then let us continue our meeting at the table' and Archie, with some difficulty, rose.

Three generous portions of caviar and one modest soup were followed by three lobsters and one piece of beef. The bottle of Chateau Margaux went down so quickly that reinforcements were summoned from the cellar. By the time they moved onto the brandy, the reopening of the Oldham plant had been thoroughly discussed, Bear's management of its intellectual property comprehensively mocked, and the launch plan for Thin-Kwik fully reviewed.

'I'm worried about the cost of some of Tiffany's plans'.

'Ken, you leave Tiffany to me. We've got to think big. We're not talking chickenfeed here'. Archie laughed at his own joke. 'I've personally reviewed all her plans in great detail' (Intimately would have been more accurate). 'We've spent, er, a lot of time together after hours working on this and Tiffany knows what she's doing'.

'But do we really need to hire the Royal Yacht Britannia?'

'As the blessed Maggie insists that the Royal Family pay its way along with the rest of us, this is one of the benefits and as a taxpayer, I intend to take full advantage of that benefit. Britannia's going to be anchored in the Thames for a week of corporate hospitality this coming March. We'll be right up there with all the blue chip companies. Tiff had to pull a lot of strings to make the booking. Maxwell wanted the same evening. He can be very abusive, you know. We need that yacht. Tiff's promised that every City bigwig will be there. We might even invite Maxwell! He could do with losing weight. If this launch goes well, Pennine shares will rocket. All those City analysts say so'.

'Archie, they don't say so to me, and as your accountant I feel I should be more involved in these meetings you have in London'.

'Tiff and me can handle them. Just you concentrate on the figures here. She's also angling to have some Royalty thrown in as part of the package. I can see the pictures now. Me and Princess Diana on every front cover. She's perfect for Thin-Kwik'. Archie's face adopted a far away look as he pondered the triumphant evening to come. His wife Mavis was already planning her ensemble, and had set, with her dietician, her target weight for next March.

'But Tiffany isn't cheap.. And all those trips to London you've been making. We've already spent half of this year's travel budget, with only two month's gone'. Ken tried to bring Archie back down to Planet Rochdale.

'I said. You leave this to me. We're talking about the biggest launch Pennine's ever had, and you're quibbling about trivial expenses. If I'm meeting top City analysts, investment bankers, gnomes from Zurich and the like, I've got to stay at the Ritz. They won't take us seriously, otherwise. Tiff says so, and she knows, you know'.

'Archie's right', said Les, with that sycophant's timing that he had honed to perfection. 'This is exactly how a start-up biotech company would do it, and Pennine's in that league now. It's all about building confidence'.

'It's all about making a profit', countered Ken.

'Now, now Ken, Les's absolutely spot-on. Confidence is the thing. I am Pennine, and these fellows want to meet me. Tiff says so. 'I've got to create the right impression. So the Ritz it is. There is one last point we've got to cover. My next meeting with the gnomes. Les, I need a boffin'.

'What?'

'They want to know what's behind Thin-Kwik. You know, how it works, results of experiments, etc. I can't do that, so I'll need to you come to the next meeting to answer their questions'.

'At the Ritz?'

'Of course'.

'I'll be there, armed with my Presentation'.

'Before we go, there is one other important procedural matter', said Terry Sharpe, in a lawyerly way, to remind them of his presence.

'What's that, Terry? Is it going to cost me?'

'It might, if you don't take more care. Confidentiality'.

'Why do you think I've spent a fortune on this deluxe lunch in a private room in the most expensive restaurant in Lancashire? If this isn't confidential, I don't know what is!'

'But the next few months are critical. We have a trademark application under way for Thin-Kwik, and that's an open process, but we don't want to draw undue attention to it. We shouldn't keep referring to 'Thin-Kwik' all the time the way we do'.

'He has a point, Archie. Remember Wiseman'. Les Fyfe never forgot Wiseman.

'Happen he does. I take it you have something to propose, Mr Lawyer'.

'As a matter of fact, I do. We should have a Code Name for this project'.

'Code Name? I like that idea, said Archie, who was happy to do anything that would put Mrs Ramsbottom off the scent of his latest enthusiasm. 'What do you propose?'

Terry Sharpe had not got quite that far in his thinking, and sought inspiration. He looked around the room, aware that the others were watching him, expectantly.

'C'mon then, Mr Lawyer, you're the wordsmith round here. What's the Code Name?'.

Terry's eye fell upon the saucer that sat on the table before him. He read the inscription: '*Qu'ils mangent de la brioche*. Perfect! Of course! In his excitement, he said it aloud.

'Eh, what's that?' enquired Archie.

'*Qu'ils mangent de la brioche*', spoke Les, the scholar. 'It's French, the motto of the restaurant. They've got it on everything. Look at the cups, the glasses, the spoons'.

'What's that got to do with anything?' Archie was growing impatient.

'That's our Code Name for the project', said Terry.

'Very inspired', agreed Les.

'Not bloody likely!. I can't be doing with Code Names I can neither understand nor pronounce. You'll have to do better than that!'

'I didn't mean in French. We'll use the English translation'.

'Let them eat cake'. Les gave further evidence that he was the intellectual of this gathering.

'I like that! I like that!' A broad smile spread across the layers of Archie's face. 'Let them eat cake. For something that makes you lose weight. Very good, Terry. Very good'. Terry received a congratulatory slap on the back. 'Where would we be without you?'

'And in the *Salle Trianon* too! *Le mot juste*, if ever there was'. Terry and Les burst out laughing.

'Eh, what's so funny? Share the joke!' But suddenly Archie did not feel like joining in. His face fell. Gaston had presented him with the bill.

Ms Tiffany Beale had been engaged by Archie Ramsbottom to handle the PR associated with the launch of Thin-Kwik. She was renowned in the business, and her name was synonymous with many of the Great PR triumphs of the 1980s. Shoulder pads, personal planners and Joan Collins had all owed a large part of their success to the ministrations of La Beale. Archie had been put in touch with her by a local vodka manufacturer, whose product she had brought to worldwide attention. Tiffany's success at persuading a significant proportion of the world's drinkers that Lancashire vodka was as good as anything Russian had been one of the legendary marketing coups of its day. As a result, she now had the status of PR guru, with the stratospheric fees and exotic fringe benefits that went with it.

Tiffany Beale had taken to Archie from their first meeting. His unpolished and decidedly heterosexual Northern character had proved a refreshing antidote to the mannered, metropolitan and largely gay male circles that were her normal habitat. What with her frustration and his libido, one thing led to another. Mavis Ramsbottom suspected nothing, and was quite relieved to have Archie otherwise occupied. Yes, she was delighted for Archie to spend more time in London discussing his new 'Let them eat cake' project. Probably some new formula for cattle cake, she thought, unaware that her husband's company no longer 'did' cattle. Mavis's settled policy for many years had been to take not the slightest interest in Pennine or its products. As long as her expensive clothes, health spa , hairdresser and lunch requirements were covered, she did not really mind which of Pennine's range of products did so.

Chapter 8: A Montecito Luncheon Party

It was a warm, sunny day at Elysium, the palatial Montecito home of Mrs Franklin J Todd IV, who for once was looking forward to welcoming her three dearest friends to her regular Thursday luncheon party, followed by bridge. She had a sensational revelation to make that would lend a proper sense of proportion to the vulgar publicity that the Borman Clinic, the pet charity of one of her luncheon guests, had been receiving recently. One month ago, during her visit to the Todd Institute for Moleculetics, she had hidden two jars of Compound X in her capacious Kelly handbag. One of these she had secreted in her medicine cupboard, which nobody, not even Mitch, her devoted personal maid, was allowed to open. The other she had given to Mitch, with instructions to feed a carefully prescribed amount daily to Missy, her husband's portly pet labrador. During the past few weeks, Prunella Drake had, with considerable satisfaction, observed Missy grow steadily thinner; a fact confirmed by the results of the weekly weigh-in that Mitch reported. She had also noted that Missy retained her famously sunny personality and gave every appearance of being a healthy, well-balanced dog. In other words, Compound X worked, seemed to have no side-effects, and the next phase of Prunella Todd's clinical trial, on herself, could now begin.

Her original plan had been to wait until her own weight loss became noticeable before telling her bridge companions about Compound X, but Shelley Borman had been so insufferable about the coverage of the Borman Clinic for Substance Abuse on CBS's weekly news magazine, '60 Minutes', that the time had come to make them aware of the sensational research going on at the Todd. That would take the shine off Shelley's recent interview with Barbara Walters. 'We've all heard quite enough about that', Prunella thought. 'And I know just how to change the subject'. Shelley Borman was naturally dumpy, ate far too much, and had failed with every diet known to medicine.

'She's been warned that if she has any more liposuction they'll be vacuuming up her vital organs. She'll be green with envy when I tell her about Compound X'.

Prunella rang for her personal maid..

Mitch appeared almost instantly. She had been lurking. 'You rang, Miss?'

'Yes Mitch. Is everything ready ?'

'Yes, Miss Prunella. Desmond has set the table for luncheon on the south terrace, and the table for bridge in the Italian Garden, as you requested'.

'Awnings in place? The sun's very strong today'.

'Of course'. Over the years, Prunella ignored much of the advice proffered by Mitch, but not when it came to protecting her skin form the sun. As a result, her flawless peaches and cream complexion belied her advancing years. The same could not be said for her bridge companions.

'I suppose Betsy Lee will complain, as usual. She loves the sun'.

'It's too late for her, anyway. Her skin's like crocodile leather. You could make her into handbags'. Mitch never lost an opportunity to restate her views on the adverse dermatological effects of exposure to the sun.

'Mitch. That's no way to speak about my friends. Now, listen'.

Mitch rolled her eyes, but listened. She sensed that Instructions were about to be given.

'Mitch, I want you to bring Missy onto the terrace when we're at lunch. I want to let them see how thin she's become since we put her on Compound X. It's a miracle, don't you think?'

Mitch blushed a deep pink and pretended to busy herself laying out her mistress's clothes. 'Yes, Miss, a miracle'.

'The cream linen today, I think'.

'And this evening?'

'For the museum benefit? Oh, why don't you put out a couple of cocktail dresses and I'll make up my mind later. Remember, I'll have to leave at 5.30 sharp. We have a Committee Meeting before the benefit begins'.

'Yes Miss, the car will be ready'.

'Good. Now help me into my dress. They'll be here shortly.'

Mitch, with some effort, zipped her mistress into her outfit. It was at least one size too small, but even she hesitated to point out that Prunella Todd was gaining weight.

'One more thing, Mitch. I'm sure Muffy Lincoln's had surgery on her eyelids. There's something different since she came back from that mystery trip to New York. I'm sure that's it. I'd welcome your opinion. Have good look at her today'.

Betsy Lee, Muffy Lincoln and Shelley Borman were among Prunella Todd's 'oldest and dearest' friends in Santa

Barbara, but none of them saw friendship as an obstacle to intense rivalry on their pet projects. This rivalry ensured that the Lee Foundation for Developing Countries, the Lincoln Museum of Modern Art, and the Borman Clinic for Substance Abuse prospered along with the Todd Institute at Bear State. African orphans, starving artists and alcoholic movie stars all had reason to thank these 'ladies who lunch'.

Betsy, darling, Muffy, darling and Shelley, darling, were all greeted warmly and examined closely by their hostess, especially Muffy, darling. Prunella and Mitch would compare notes afterwards. It was their favourite diversion at these weekly gatherings. Any new dress, new jewellery, weight loss, weight gain, change in hair colour, scars, facial bruising or new bandages exhibited by luncheon guests would be critically reviewed by the two leading ladies of Elysium ere bedtime.

Luncheon over, the four ladies were sipping their coffee and Prunella was beginning to think she would strangle Shelly Borman if she mentioned Barbara Walters one more time, when Mitch casually strolled past the terrace with Missy in tow.

'Look, there's Mitch!' cried Prunella, excitedly, as if surprised. The other three women looked at her and then at each other. Of course it was Mitch. They could all see that, but they had never before seen their hostess become excited at the appearance of her personal maid. Irritated, yes. Intimidated, often. Relieved, frequently. But excited, never. It got more bizarre. Prunella stood up.

'Look, Missy's with her. Missy, darling, come and say hello to mummy'.

This was too much for Mitch, who had to stifle a threatening laugh. Meaningful glances were exchanged by the three guests… 'Mummy?'… Missy, who had never before been favoured by Prunella, was nonetheless grateful for the attention and hoping for a titbit, scampered towards the luncheon table, tail-a-wagging.

'My, look how thin you've grown. So lean, so sleek, so fit. Isn't she wonderful, girls?'

'Gosh, she has lost weight', said Betsy Lee, who had labradors herself and was fond of Missy 'How did you do it?' At this point, mission accomplished, Mitch headed across the lawn at speed. She had other work to do, using Mr Todd's high-powered binoculars, which she had secreted in her pocket. Mitch took up position in the shade of a pergola just beyond the fountain. This would, she had calculated, provide her with excellent cover from which to observe Mrs Lincoln under powerful magnification.

'New treatment', purred Prunella.

'We must try it on our two. Which veterinarian? Dr Knowles?'

'No, no, Betsy dear', said Prunella with a superior chuckle. This is not available from any vet. It's been developed by our wonderful scientists at The Todd. It's a major breakthrough in weight control. Papers have been published in the leading scholarly journals. Franklin and I are very excited'.

'And you're trying it out on Frank's dog? Isn't that risky?'

'Actually, it has been fully tested', asserted Prunella rather defensively, knowing this to be untrue.

'And Frank's happy to use it on Missy?'

Prunella evaded the question. 'She was getting very fat. Something had to be done. It's very benign. Looks just like sugar'. She was becoming uneasy with the direction of this interrogation, but was saved by Shelley Borman.

'Who cares about dogs! What about people; does it work on people?'

'Well, of course, it's early days, but the first results are very promising'. More lies, but Prunella wanted to get off the subject of labradors. 'But you have to remember that it will take at least ten years before it can be made available for people. Extensive testing needed: lots of regulations, you know'.

'Ten years!' exclaimed Shelley Borman. 'I'll be, I'll be …almost sixty five'.

The other ladies exchanged more meaningful glances. Prunella thought she would show off her recently acquired knowledge of neurochemistry. A mini lecture ensued on hunger centres in the brain, their stimulation and suppression and how the latter was the key to easy weight control. Her guests were transfixed. None more so than Shelley Borman, who asked a torrent of questions.

'How quickly does it work?'

'When will it be on the market?'

'Isn't this too good to be true?'

'How big is the dose?'

Eventually, after much badgering, Prunella agreed to show them her supply of Compound X once they had finished their bridge game.

That afternoon, Shelley Borman, usually the sharpest player of the four, proved a most unsatisfactory partner for Prunella. She misread cues, forgot hands and seemed generally distracted. By the time tea arrived at 4.30, signalling the end of the game, Prunella was at the end of her tether.

'Shelley, you didn't concentrate at all today. What's wrong with you?'

'Oh, I expect it's my antimalarials', she lied. 'We're off

to South Africa next week, you know. I've started taking them'.

'Poor Shelley. You should have said. Would you like to lie down for half an hour?' Prunella's concern was touching.

'No, don't trouble yourself on my account. I'll be fine'.

'It's no trouble. You can rest on my daybed upstairs. I have to go up anyway, to get the stuff'.

'The stuff?' Shelley paled immediately. 'Oh? Of course, the stuff. Well, if you're going upstairs anyway, maybe I'll lie down for half an hour'.

After a brief struggle, Shelley Borman pulled herself to her feet, staggered slightly and rested a hand on Prunella's shoulder as she made a halting progress into the house and upstairs to the bedroom. 'Here we are. Look, you lie down on this daybed. That's better. Can I get you an aspirin or anything?'

'No, thank you, I'll be fine'. Shelley laid her head on the pillow while Prunella carefully arranged a cashmere throw over her. Shelley closed her eyes, or at least appeared to.

In truth, Shelley Borman's eyes were only half closed and she was able to observe her hostess open the door to her bathroom, emerge holding a white, plastic jar, tiptoe across the room and gently close the door to the hall. This process was repeated, in reverse, some 10 minutes later, by which time Shelley Borman gave every appearance of being in a deep sleep. She heard every word of Prunella's whispered conversation with Mitch as they stood over her.

'Poor thing. These antimalarials can have such awful side-effects'.

'So I've heard'. Mitch, who had seen the side-effects though a pair of high-powered binoculars, had found them unconvincing.

'Mitch, let's leave her to sleep a while. You call her house and let them know. Keep an eye on her'.

'What about the pekes? You know she doesn't like dogs.'

Oh, leave them here. The darlings are having their nap: they won't bother her. Look, I'll change in the other room. I'm running late'.

Their guest's eyes flickered briefly, as if in a disturbed dream. The flicker was sufficient to allow her to see Prunella carry the jar into her bathroom. A shower was turned on, doors opened and closed, but after 20 minutes, all was quiet. Prunella Todd had departed for her museum benefit.

Shelley Borman had, meanwhile, revived sufficiently to call to the kitchen for tea. 'I like it very sweet, so bring plenty sugar. Not lumps, mind, loose sugar'.

She was taken aback when, some minutes later, Mitch entered the bedroom with the tea tray.

'Oh, Mitch. This is very kind of you. I expected the parlour maid to bring it'. Like all who knew Mrs George Mitchell, Shelley Borman had a healthy respect for her legendary intelligence and would have preferred her to be elsewhere at this precise moment.

'Mrs Todd asked me to keep an eye on you. Are you feeling better? Your attack was very sudden. I brought some biscuits in case you feel hungry'.

'Did you bring sugar?'

'Yes, ma'am. White, loose, as you requested. Would you like me to remove the dogs?' Mitch watched her intently.

'The dogs? 'Oh, no, no … no, leave them here. They're actually very comforting'.

Mitch looked askance at the four snoring pekes, each in its own dog bed. She turned and made her way to the door.

'Very comforting? She can't abide dogs. But then I didn't think she took sugar in tea either. A very contrary woman. What is she up to?

'Is there anything else I can do for you, ma'am'?'

'Oh, yes, can you ask my driver to pick me up in 30 minutes? I'll be ready to go home by then'.

'Very well, ma'am'.

Shelley, sensing that Mitch was on high alert, waited for a good 10 minutes before tip-toeing to the bathroom, bowl of white, loose sugar in one hand, Gucci purse in the other. Once in the bathroom she made straight for the large, mirror-fronted cupboard beside the sink. She quickly found the jar, emptied its contents into her purse and refilled it with sugar. She replaced the jar in the cupboard, flushed the toilet, splashed water in the sink and dried her hands. Her return progress across the bedroom took her first to the tea tray and then to the slumbering pekes. Shelley Borman needed to cover her tracks and was quite prepared to incriminate four entirely innocent small dogs to do so.

'Little doggies, look what Aunt Shelley has got for you' she whispered sweetly.

The scent of biscuits being waved in front of their snouts woke them instantly, and they scampered excitedly towards their new benefactress. Montecito's leading canophobe recoiled from their snuffling advances and threw the biscuits onto the floor. While the dogs were noisily devouring this unexpected treat Shelley scattered the remaining sugar around their beds: enough to convince even Mitch of their guilt.

Shelley Borman was satisfied with her handiwork. 'Well, well little doggies, what will your mistress have to say about this?'

Four little black muzzles, covered in sugar and biscuit

crumbs, looked up at her, pleading for more.

'Guilty. Guilty. Guilty. Guilty' she hissed, pointing an accusing finger at each in turn.

Now she was on a high. She not only had her prize, but she had incriminated Prunella's loathsome dogs. She could contain herself no longer.

'I'm to be beautiful and thin! And you'll be in the doghouse! Ha, Ha, Ha!'

Her cackle, as she slumped back onto the day bed, was worthy of Cruella de Vil.

Then Shelley Borman came to her senses and checked her watch. Mitch would call her in 10 minutes, so she would be downstairs in the hall in five. After calming the pekes, which were now on the lookout for more food, she fixed her face, brushed her hair and smoothed her dress. Picking up her now bulging purse, she made a detour to the tea tray, where she upset the empty sugar bowl and biscuit plate, before opening the door and blowing a parting kiss to the pekes. She met Mitch at the bottom of the stairs.

'Feeling better, Mrs Borman? I was just on my way to collect you.'

'Much better, Mitch. Thank you'.

'The car's already waiting for you. By the way, did you leave any biscuits? The dogs, you know. Always on the lookout for food'.

'Biscuits? Oh, no … I mean … yes. I didn't touch them. I wasn't hungry'.

'Those dogs. I'd better get up there.' Mitch lifted her skirts, ready to sprint upstairs. 'Goodbye, Mrs Borman. Desmond will see you to your car'.

'Oh, Mitch. I suddenly feel a bit faint. Could you help me to the car?'

Mitch had no other option. 'Very well. Ma'am. Take

my arm'.

As soon as Shelley Borman was safely deposited in her car, Mitch dashed back into the house and up the stairs. As the car pulled away, its passenger clearly heard Mitch's raised voice, followed by the nervous yapping of small dogs. She sank back into her seat and patted her bulging purse, which sat securely on her lap.

Chapter 9: Klomp Wins a Prize

Henri de Klompenmaker sat at his desk in his London office. A rare smile lit his face, he was courteous to his secretary; his office door was open to all; he exchanged pleasantries with his colleagues and he had returned with chocolates to be shared. The reason for this uncharacteristic bonhomie? Jet-lag, perhaps? A substantial salary increase? Had he found the postage stamp of his dreams at a bargain price? Had he been bitten by something? There was, among the members of his department, a mild curiosity about this odd behaviour from the usually unsmiling, taciturn and closed-door Klomp. But only mild. His was not the type of personality that engaged the interest of his colleagues for long, and none of them wanted to volunteer for a direct conversation with him to find out. All could recall how any discussion with him ultimately veered in the direction of philately, a long and boring journey, and none had the stamina this morning to face it. Instead, they would wait until Miel Flick, the resident Management Trainee, arrived.

In the short time she had been among them, Miel Flick had made Klomp her particular study and could without doubt be described as London's leading Klompenologist. It was widely suspected, however, that her motivation for this study was ambition rather than a fascination with the personality of Henri de Klompenmaker *per se*. Like many of her ilk, she was a power groupie, and knew that Klomp, currently held in such high regard by The Chairman, could be useful to her. Whatever her motives, Miel Flick would find out what lay behind his sunny mood this morning. She would then delight in delivering the news to all in that rapid, telegram delivery that she had probably learnt at business school, and which had earned her the nickname MBC – Miel Broadcasting Corporation. Yes, MBC would save them all a lot of effort.

Miel Flick was one of the expanding flock of Management Trainees that had been the fruit of one of Nigel Archer's earliest Executive Orders. He had decreed that every Director must have one, all part of his 'shaking up' exercise. Klomp might ignore his colleagues, but never an Executive Order, and he appointed a Management Trainee as soon as he became a Director. He needed a status symbol that would reinforce his prestige in the organization. This was important, as the other status symbols he had worked so diligently to acquire had quickly become threadbare. His staff did not appear to be sufficiently intimidated by his Director title, nor could they watch in awe as he drove to work in his new BMW. Almost nightly break-ins outside his residence in Mafeking Avenue had forced him to give it up within a month of receiving the keys. His procession of Management Trainees had proved much more of a success. This one was large, blonde and mouthy. Nobody could ignore Miel Flick.

She also had practical value as a confidante-cum-spy. Klomp had singularly failed to open channels of communication with his London colleagues since moving there. Few of them appeared to wish to spend any time with him. Nor did he seek their society but he did want to know what was going on, and was convinced that he was not being told. From his fellow directors elsewhere in the organization, he had learned how effective the Management Trainees could be, as spies. Like parrots, they were smart, noticed everything and repeated what they heard to chosen individuals. They also had a magpie instinct to gather gems of information. Unlike swans, they formed no lasting attachment to any individual or location. Birds of passage, their object was to move around the company as much as possible, pick up a superficial knowledge of every function and progress up the career ladder. As a species they were wily enough to ingratiate themselves with their Director, as this would oil the wheels of career progress The key to ingratiation was information. Information on colleagues, projects, other departments in which they had spent a term. The more they shared with their Director, the more access they had. The more access they had, the more influence they had. The more influence, the more power. The more power, the more sought after. The more sought after, the better the career prospects. The more career prospects, the more money. Miel Flick understood little of publishing, but grasped these fundamentals very well. In this way, she had acquired a very good 'press' in the upper echelons of the SI organization, had been voted Management Trainee of the Year, and understood that her name had been favourably 'mentioned' to The Chairman himself. This meant the ultimate prize, a position in 'Corporate Strategic Planning', could be hers if she continued to play her cards well.

Miel Flick was a textbook example of the Management Trainee. Ambitious, vocal, energetic, devoid of principle and possessed of a mind that was strong on analysis but free of imagination. She had little interest in publishing, but was very interested in being part of a large, successful corporation through which she could chart a path to the top, while accumulating fringe benefits along the way, By the end of her first year with SI, she had already worked in three different locations and had sampled (she would claim 'mastered') six different jobs. Her knowledge of SI was thus impressively broad, if dangerously shallow. She could speak for five minutes on any aspect of the business, but would run out of things to say after ten. She had worked for Henri de Klompenmaker for four months. Theirs was not a relationship based on mutual respect, but both saw that the other had something to offer. She was a useful accessory to him, while he was a useful stepping stone to her.

As was her habit, Miel arrived at the office at 10.00 am. She took full advantage of the flexible working hours at SI and found it suited her to start late. This, it was widely assumed, was because starting late implied that she finished late and, it was surmised, allowed her maximum overlap in hours with the all-important New York office, 5-hours behind London. Miel revelled in this little conceit, as she realized how much it helped her 'corporate profile' to allow her colleagues to gain the impression that she was in daily contact with New York and that there were, her colleagues would assume, important, presumably top secret projects that required her personal input and daily conference calls. 'Time zone management' was not something formally taught at business school, but Miel and her fellow trainees had all absorbed its core principles. These were: First, always work for a company with offices spread over several time zones. Second, ensure that one's involvement in projects is well spread across these time zones. This will require frequent business trips to the other locations, daily phone calls and will create an appropriate level of mystique with one's colleagues in the office that happens to be one's base at that moment. Proper time zone management also requires that the practitioner must move frequently between departments and offices. This not only provides a ready excuse as to why one has little time for work that shows any local results, it is also essential to survival, as the inevitable questions about these projects grow with time. Fortunately, the SI approach to trainees ensures optimal 'Time Zone Management'. The firm rule is no more than six months in one function, in one location.

Immediately upon her 10.00 a.m. arrival, with a predictability that Klomp admired, Miel Flick entered his office for her morning briefing, her antennae finely tuned to pick up the slightest corporate nuance.

Today's would take some time, as Henri de Klompenmaker would have to relate, with that attention to detail for which he was renowned, the exact circumstances of his triumph in Bermuda, while Miel Flick would have to bring him up to date with the many important developments that had taken place in the SI global empire. She had much to impart, but the usual volume of material would be overwhelmed by one particular item of news which had to be announced in the largest and boldest of tabloid headlines. For the last two days, since hearing this news from her contact in The Chairman's office, she had been giving careful thought to the wording she would use in her announcement. Miel had, by now, arrived at an appropriate, 2-word show-stopper, but as she passed Klomp his coffee, she knew that etiquette demanded that she permit the Director to relate his tale first. He was clearly eager so to do.

'They were stunned', he declared, 'I think it was the first time any of them had been made to consider proper, quantitative measures of journal performance'.

'Bermuda went well, then?' queried Miel.

'Couldn't have gone better. I think I opened up a whole new management universe for them, at least for those who understood what I was saying'. He described his presentation in detail, which was rather unnecessary as Miel had written most of it. But she forbore to remind him of this and smiled patiently as he took her through the story of 'A Quantitative Approach to Journal Management'. As one would expect from a publisher, it had been heavily edited.

He was sure that Miel would not wish to know about the fiasco that the session at which he spoke had almost become. When it opened that evening, apart from the three scheduled speakers, there were only two people in the audience and nobody in the Chair to supervise the proceedings. It happened that 50% of the audience was Lionel Grove, who made a point, on occasions such as this, of attending the performance of any SI speaker and saw it as his duty, as Director of Corporate Communications, to review that performance at length in his report to The Chairman. Careers had been made or broken on the strength of Lionel's reviews. As he usually understood little of the content, he would concentrate on style of presentation, appearance of speaker and, most importantly, on audience reaction. This was exactly what Nigel Archer wanted. He liked to know who it was safe to send out to represent the company in public, and many had been denied further exposure following a negative review from Lionel.

By the time the second speaker rose, the audience had grown to three, with one already asleep. Company pride demanded that there should be more than three spectators for a presentation by an SI Director. Lionel knew drastic action was required and did not hesitate. Slipping out of the room, he headed for the bar, where he suspected the missing audience might be found. His instinct served him well. They were all there, regaling one another with tales of deep-sea fishing, scuba diving, sailing and the other adventures they had enjoyed on that perfect, sun-filled afternoon. Lionel was greeted warmly by the pack of baying publishers as he strode into their midst. Marlene was one of the first to give tongue.

'Lionel, *quel surprise*! Why are you not at the session? I thought one of your people was speaking?'

'If you mean Henri, yes. What are you all doing in here? They're speaking to an empty room over there'.

'*Quel domage*, but what do you expect if you will arrange a session on 'Quantitative Management Tools in Publishing' immediately before dinner on the first evening!' retorted Marlene.

'Actually, it's very interesting and I think you would all learn something'. This, he knew, was not a compelling argument for this crowd and was unlikely to persuade them to part with their cocktails. Stronger medicine would be needed. Lionel took Marlene aside.

'Look here, ducky, you know how The Chairman is always asking what we get for all the sponsorship money SI gives this retreat every year. I believe it is around 25% of the total' (He actually knew precisely, to the penny) 'And I suppose you will be expecting something similar for the one you are organizing in Switzerland next year'.

He had to say no more. Marlene caught his drift immediately. She was well aware that Lionel, as SI's Director of Sponsorship and Public Relations, had enormous influence over this budget, and that a negative report from him on this gathering would give the cost-cutting Archer the excuse he was looking for to give it the chop. Marlene's arrangements for Switzerland were both well-advanced and lavish. They would not survive a 25% reduction in sponsorship and her reputation as one of the publishing world's greatest party givers would suffer a mortal blow. She sprang into action, and worked the room as only she knew how. A word here, a whisper there were sufficient for a crowd that knew, if they knew anything, the side on which their bread was buttered, and also who was providing the butter. A general exodus to the lecture room was under way within minutes. By the time the third speaker rose, there was standing room only. Lionel Grove surveyed the scene with satisfaction. The third speaker was Henri de Klompenmaker.

'Miel, when I rose to speak, the session chairman had to ask for extra seats to be brought in; it was the only well-attended session at the whole conference'. Klomp was still untutored in the ways of the Newtonian Academy of Scholarly Publishing. 'There were literally gasps of excitement when I demonstrated, with statistics alone, the huge improvement in the position of *Transactions*.' (Marlene had regretted urging the audience to 'show a little enthusiasm for once). 'I got a standing ovation at the end, and one or two people were even shouting bravo!' Klomp had no inkling that this enthusiasm, while loudly expressed in the course of his presentation, had been prompted by the more distant, but suddenly endangered, prospect of the Swiss Alps in March next year. 'There was a lively debate at the end, and people wanted to know more about our methodology'.

Again, his editorial hand was at work. Klomp had not known that he was safe only as long as Lionel remained in the room. When the question and answer session ran on too long for his liking – he had to check on the arrangements for that evening's dinner – Lionel left. This did not go un-noticed by the eagle-eyed Marlene, who had been a model of good behaviour until that point, but now decided to have bit of sport with Klomp. He broke into a sweat as soon as she rose; having deluded himself that it was his intellect rather than Lionel's presence that had kept her silent.

'Henri, you have spoken a lot about the citation statistics for these papers published in *Transactions*, but have said nothing about the research actually being reported in them. It's obviously exciting. I'm intrigued. What is it about?'

'It's chemistry', replied Klomp.

'I think we had all sort of assumed that, *Transactions* being a chemistry journal. Could you be more specific? This work sounds very exciting'.

'No, I'm not a scientist. You can read the papers for yourself'.

'Time was when the management of SI took an interest in what they were publishing and even understood it. Sir Algernon Brogue was one of the great publishers of his generation.' At this murmurs of approval rippled round the room. Sir Algernon had been admired by all. Marlene beamed.

Klomp scowled. 'That's micromanagement', replied Klomp. 'I take a more strategic view'.

'I suppose that's one way of describing it. Let me ask you something you do know about, then. What about these statistical tools you describe? Your point is that the number of citations a paper receives is a measure of its quality'.

'Generally, yes'.

'But surely, very bad papers can also be highly cited'.

'They are not statistically significant for a journal that publishes as many papers as *Transactions*'.

'My point still stands. The number of citations a paper receives is not necessarily a measure of quality. Things can be cited, shall we say, for the wrong reasons'.

Marlene was impressed by her own tenacity on this

point, as were her audience. But she felt strongly. Klomp and his like, in her view, were trying to reduce publishing, which to her was an emotional as well as an intellectual process, to a collection of numbers to be added, subtracted and otherwise manipulated. She went on: 'So what you are saying is "Any citation is good citation"'.

'Your words, not mine', replied Klomp, hoping the Chair would step in to rescue him. He did not.

'To rely entirely on citation statistics to lead your editorial policy seems both shallow and misguided. Where does professional judgement come in?'

'You mean gut instinct. Not a good basis for directing a business'.

By this stage, the audience, many of whom had built considerable fortunes in publishing on the basis of gut instinct, shifted uneasily in their seats. Having been driven reluctantly from their pre-dinner cocktails, they were becoming concerned for the wellbeing of their digestive systems. Dinner beckoned. It was known that Marlene and Klomp had crossed swords at the London Forum. She had relived this contretemps at length in the bar and they were now being treated to an action replay. But the appetite of the audience was for dinner, rather than for further debate and a silent consensus began to emerge that this discussion should now be terminated. This collective wish was communicated to Marlene by a sharp kick to the shins from her neighbour. Always sensitive to her surroundings, she did not rise again to Klomp's bait.

Klomp was grateful for this, as he feared a repeat of his humiliation at the London Forum. The Chair was grateful, as he was becoming anxious about how he was going to bring this increasingly testy exchange to an end. Above all, the audience was grateful, as they could now proceed to eat. Their heart was with Marlene, but their stomach was with Lionel, and he would be waiting to marshal them for dinner.

As they strolled across the lawn to the moonlit terrace on which they were to dine, Marlene was left with a nagging curiosity. She must find out exactly what these papers were and made a mental note to have her secretary obtain copies of them. The others were left with the general impression that there might be something in this statistical stuff, and would get their boffins to look into it when they got back home. Klomp was left with the impression that he had had something of a triumph; the crowd, the gasps, the applause. Lionel, who had not been present for the questions, would certainly have left with the impression that he had had a great triumph and, more importantly, would convey this impression to Nigel Archer in his report. Klomp would be able to relax in the knowledge that there would be no further serious discussion over dinner, where the accepted practice seemed to be to avoid, at all costs, any discussion of the substance of the presentations.

Klomp gave Miel an easy opening as he finished his tale, from which he had erased the debate with Marlene Pym. 'And no other speaker had a capacity audience all week', he told her with glee. 'It looks like I am going to be invited back to give a follow up next year. So what's been going on here?'

Miel's moment had arrived. 'The Snurt'.

'The Snurt!' he exclaimed. The mere mention of its name was enough to drive thoughts of easy triumphs in Bermuda from his mind. 'Of course it's that time of year again. Who?'

'You!'

'Never!'

'Yes, you, Henri. I got the news from my source in The Chairman's office'.

Now it was Klomp's turn to be speechless. He stared open-mouthed at Miel, whose rather frighteningly broad-toothed grin gave the perfect orthodontic confirmation that she was being serious. Miel never joked about stuff like this. 'The Chairman will be calling you personally with the news. You will receive the award at the Annual Strategy Meeting next month'. She left him alone to enjoy the moment. No need for more news today.

'The Snurt!' he thought. 'Well, I do deserve it, of course. But for what?'

In the self-contained universe of SI the Snurt Award for Administrative Excellence was one of the most sought after of all the company's prizes. It was named after Wilbur Snurt, a famously parsimonious Finance Director of the SI Defence Systems Division, whose attitude to expenditure in general and corporate entertainment in particular, had been formed in his childhood on a poor farm on the prairie. The financial tools he had developed for keeping tabs on SI Defence Systems in Chicago had so impressed the human calculating machines at SI's corporate headquarters in Lichtenstein that they had been imposed on the other SI Divisions. The Snurt was awarded annually to the individual who had made the most significant contribution towards improving SI's administrative or financial procedures. In his drive to cut costs, Nigel Archer had designated the Snurt the most important of all the SI awards, had doubled its monetary value and had made it the centrepiece of the annual Corporate Strategy Meeting Dinner. The management of the SI Publishing division had traditionally taken a disdainful view of the Snurt and took pride in the fact that none of their number had ever won an award that they regarded for the 'bean-counters'. Henri de Klompenmaker, though they did not yet know it, had blemished this perfect record.

That afternoon the phone rang and his secretary announced, with due solemnity, 'The Chairman's Office', and connected him to Nigel Archer's PA.

'Mr de Klompenmaker? The Chairman wants to speak with you. Can you hold?'

'Of course', replied Klomp, while unconsciously straightening his tie and adjusting his posture. Should he stand up to take this call? But his deliberations were caught short by a bark at the other end of the phone.

'Henri? De Klompenmaker?'

'Yes, Chairman'.

'Good! Look here. We've decided to award you the Snurt this year. What do you think of that?'

Klomp, still basking in the warm glow of Miel's earlier revelation, would have had no problem transmitting his delight, had Nigel Archer given him a chance to do so. 'Well, Chairman, naturally this comes as a great …'.

'Surprised, eh? Well so were the Board when I suggested it. Never been awarded to anyone in the Publishing Division before, they said. The Publishers get enough awards already, they said. Could harm morale in the other Divisions, they said. Important to acknowledge the backroom boys, they said. Pish, I said, we want to encourage efficiency and cost control in all parts of the business. You've shown the way in publishing and I want to show the others that you are an example to follow. Don't you agree?'

'Well, Chairman …'.

'Now listen. You've got to keep completely oyster on this'.

'Completely what?'

'You must not talk to anyone. The name of the winner

is never revealed until the award is made at the Corporate Strategy Meeting Dinner'.

'Of course I know that!' Klomp thought loudly, to himself. Anyone steeped in the Snurt legend as he had been was well aware of the rules of the game. How well he recalled the excitement in the Elevator Division in the weeks leading up to the announcement of the Award. The lists of invitations to the Dinner were eagerly sought and poured over. Bets were placed and red herrings were laid. People in the finance department actually spoke to one another. Of course he knew all about the Snurt!

'Naturally, I'll say nothing, but …'.

'Good and remember you'll have to make an acceptance speech, with a slide show. Congratulations. Goodbye!'

Klomp was left holding the phone and thinking about The Speech. He would need to work on that. The whole point of all this notice was to give the winner an opportunity to prepare a speech that would be published in the company in-house magazine *SI Universe*. To match the orations of previous years it would have to be pompous, pedantic and prolonged. Had his colleagues been asked, they would have pointed out that Klomp could easily achieve these objectives with no preparation whatsoever, but he was less confident of his talents. This speech would be his manifesto, on record in *SI Universe*, along with photographs of him in relaxed poses and descriptions of his hobbies. What part of his extensive stamp collection would he select for the paper? He had much to think about.

He recalled, with awe, previous Snurt winners, as well as the titles of their Acceptance Speeches. There was Günter Strauss, Agricultural Machinery Warehouse Director, who had entertained them for 50 minutes with 'Twenty years in the Warehouse'. Who could forget Anna Maria Ickx, Deputy

Director of Global Exhibitions for SI Defence Systems and her 90-minute tour of four continents in 'Conferences I Have Known' (this had been accompanied by a particularly colourful slide show). Martin Smith, Head of IT in Shipbuilding, had kept them bemused for over an hour on 'Bytes Don't Bite'. Klomp was well aware of the size of the challenge and to meet it, he realized that he would have to devote most of his time in the coming weeks to preparing his speech and slide show. But first he would have to speak to Miel. This was not difficult, as she was loitering with intent outside his door. Her 'contact' had informed her that the Chairman was on the phone.

'Look here Miel …'

She looked.

'This Snurt. You've got to keep completely oyster about it'.

'Completely what?'

'Don't mention it to anyone. It has to be kept quiet until the Dinner'.

'I know that. Who are you inviting to the Dinner? The winner is allowed to invite three close colleagues. It's a great opportunity to network'.

Klomp had forgotten about that, but he got her point. 'You, naturally, will be invited'.

'Why, thank you Henri, that's very thoughtful', purred Miel, 'I hear that Lola Santiago is speaking at the strategy meeting. It would be thrilling to see her. She is such a positive role model for women in business'.

'Indeed. The Einstein of pricing. She spoke in Bermuda, you know'.

'No? Did you hear her'.

'Naturally, and I think I was the only one in the room who understood her. You know what the NASP crowd are like'.

Miel did not know, actually, but would certainly not admit as much to Henri de Klompenmaker. She knew how important it was to be seen as an 'insider' in publishing. She changed the subject.

'Of course. By the way, let me know if I can help you with your speech'.

'Good thinking. It's going to take up a lot of my time over the next few weeks'.

'Glad to be useful', were Miel's parting words as she left the room.

'How does she do it?' mused Henri, with a mixture of fear and admiration.

He summoned his secretary, to whom he divulged very little.

'Look here ...'

His secretary looked.

'I want all my non-urgent travel cleared for the next 8 weeks. I need to be here. Chairman's Project'. His secretary would not question that. 'Bring in the diary and we'll go through it now. I need to decide what to keep and what to drop'.

His secretary read out the list of engagements.

'Corporate Audit Process Committee ...'

'Keep'.

'Government Seminar on UK Five Year Scientific Research Plan, London ...'

'Delete'.

'SI Human Resources Policy Committee, Sydney ...'

'Keep'.

'The Future of Publishing, Does the Internet Have a Role?, CERN, Geneva ...'

'Delete'.

And so they went on, until she came to 'Bear State University, Todd Institute ...'

'The what?

'It is a university in California. Professor Fyfe has arranged for you to go'.

'Delete. Send Fiona Hamilton'.

'But he went to a lot of trouble to have you invited'.

'Send Fiona Hamilton'.

'Finally, SI Corporate Strategy Meeting, Cap Ferrat'.

'Keep. That will be all', said Klomp with a smile.

'There is one more thing. Fiona Hamilton wants to see you urgently'.

'Tell her to send me a memo'.

'But, it's to report on the *Transactions* Board Meeting in Estonia'.

'Look here, that will be all', he snapped.

Now he had some private time to relish his triumph. He could tell nobody, of course. There was little point. The Snurt would mean nothing to his parents, to whom his successful business career was a complete mystery. His brother, the butcher, and his sister, the nightclub chanteuse, had shown no interest other than jealousy at his extensive travel.

During his many years in SI Finance, Klomp had been one of the most ardent disciples of the Snurt school of management. Cost-minimisation had been his byword, and his travel arrangements would have made Scrooge proud. That was until he moved to the Publishing Division. Well, he had reduced the travel budget by half; fewer people, less travel. Why should he be a martyr; he was already living in exile. Wasn't that enough of a sacrifice? After all, he was a Director, wasn't he? It would look bad for the company if Directors travelled economy. And, if he looked deep into his heart of hearts, he would have to admit that he actually enjoyed it. The irony of it, he wins the Snurt after he has succumbed to a pampered executive lifestyle of chauffeur-driven cars, business class travel and first class hotels. Was this a warning? A hint to return to the straight and narrow? A timely reminder from old Wilbur? Klomp banished such thoughts from his mind, as he looked forward to the forthcoming Strategy Meeting and Award Dinner at the Hotel Excelsior on Cap Ferrat. Somewhere he had never been, but Lionel Grove, who had made the arrangements and had told him all about it in Bermuda, assured him that it was the best hotel on the prettiest spot on the Riviera and that early summer was exactly the right time to go.

The weeks between the News and the Event flew by for Henri de Klompenmaker.

They also passed very quickly for his colleagues in the office, who had all noted his newly cheerful disposition and even a tendency towards banter that had hitherto been absent. The quickly arrived at consensus was that Bermuda had done him a power of good and that if it resulted in him behaving like a normal person, further visits there were to be encouraged. It was left at that, and London quickly adjusted itself to the Happy Henri. None guessed at the real reason.

The fact that Miel Flick was closeted with Klomp for the greater part of each morning caused no surprise. This was quite normal. Only the topics – or rather topic – discussed was unusual. Klomp was more than happy to accept Miel's offer of help to prepare his Acceptance Speech. She had an instinct, borne of her relentless networking, for those corporate buttons that should be pressed for maximum impact. She was also much more fluent in 'management speak' than he was. Miel could distinguish her strategic objectives from her strategic options and could spot a strength, weakness, opportunity or threat, at 50 paces. Yes,

Miel would keep him straight.

It took the two of them a week to decide on the topic and the approach he would take. Many flip charts were consumed in the process. Klomp's management theories had, as yet, only been tested on academic journals, but emboldened by his success in Bermuda, he thought that now was the time to propose that they be applied across the whole SI Publishing Division. Miel initially urged him to be more cautious, as she felt that the results were not yet ready to be shared too widely. But Henri de Klompenmaker was determined to build on his Bermuda triumph and she saw little to be gained by arguing with him. After all, she would be moving to another department within three months. His Snurt Award acceptance speech would be 'A Quantitative Approach to Publishing'.

The following weeks were devoted to writing The Speech, followed by several days of intense rehearsal. In the mornings in front of Miel, and in the evenings in front of a mirror at home.

The day of his departure for Cap Ferrat eventually arrived. His secretary handed him his tickets, currency, and the SI Annual Strategy Meeting Information Pack', prepared by Lionel Grove, and his dinner suit, hired from Moss Bros, which he would naturally charge as a business expense. In a change from the original plan, Miel would be flying out with him. She had informed her contact at The Chairman's Office as soon as she had extracted her invitation to the Award Dinner. As she had to travel out anyway, she offered her services as one of the 'Meeting Stewards', positions reserved exclusively for, and vigorously fought over by, the Management Trainees. Their job was to help Lionel Grove with the local arrangements, ensure speakers arrived on time with their presentations in the desired form

(Lionel was very strict about this) and check that all the participants actually attended the events to which they had been assigned. Lionel was a demanding taskmaster, but it was worth it for the opportunity it gave them to network with Senior Management.

Nigel Archer took the closest personal interest in these arrangements. This was His Show and he personally approved the list of attendants. In due course, The Chairman's Office let it be known that Miel was to be one of the Meeting Stewards, which triggered the next phase of her plan. Management Trainees were not strictly allowed to fly business class, but could do so if they were accompanying a Director. At the end of one of their speech-writing sessions, Miel mentioned to Klomp that she would be one of the Meeting Stewards and asked if he would mind if she took the same flight as him.

'Of course not', was his response. Unusually for Klomp, he had forgotten the precise procedural consequences of his permission. But he was absorbed in his speech.

Miel was only too well aware of the procedural consequences, and as soon as she left his office, proceeded to explain them in great detail to his secretary. No, she had not yet booked Henri's flights. No, he did not usually travel that airline. Yes, she would book them two seats, together, in business class. Yes, she was aware of the Travelling With a Director Rule, but had never had an opportunity to implement it, as none of Henri's staff were very insistent about being promoted to accompany him in business class.

Klomp and Miel were not the only SI people on the flight. Miel had been informed that Nigel Archer himself would be among the passengers, but to her disappointment he had changed his plans and would be flying in from Frankfurt. But the rest of the Head Office crowd was there, which had already provided her with some decent networking opportunities in the Lounge at Heathrow. Once on the plane, she had taken advantage of an empty seat next to the Head of Corporate Finance in order to continue networking with him more intimately.

Left to his own devices, Klomp opened the 'SI Annual Strategy Meeting Preliminary Information Pack' and began to read. This was his third such meeting, so he knew the form by now. The pack was beautifully (and, he suspected, expensively) produced, the work of Lionel Grove in his capacity as Director of Corporate Events. This was Lionel's biggest Event of the year, and he would ensure that things were done properly. On the thick, glossy folder a stunning Mediterranean sunset shone out. Applying his considerable powers of deduction, Klomp assumed that this must be Cap Ferrat. Opening the folder revealed the usual, clearly marked sections.

Section 1: Dress: Lionel was very strict about this. During the day, when the designated ambience was to be 'relaxed', clothes were to be 'business casual', which meant 'no jeans, no sandals, no vests, no cut-offs, no track suits, no leather pants, no tee-shirts-with-logos, no see-through tops, no baseball caps indoors, no shorts or swimming trunks – except-by-the-pool, no socks-with-logos'. The evening dress code was rather more exacting: 'On the first two evenings: jackets to be worn by gentlemen, with long-sleeved shirts; ties optional (everyone knew what Lionel meant by 'optional'; all would be wearing ties). Ladies to wear dresses or smart trousers (the last was a very reluctant recent concession to modern dress habits). The Awards Dinner is Formal. Black Tie for gentlemen and cocktail dress for Ladies'. So it continued for two pages, with strictures on what to wear, when and how.

Section 2: *Meeting Timetable*: This was the most complex part of the package. It began with the injunction 'The Timetable Will Be Strictly Adhered To'. This was an essential requirement, given the large number of parallel sessions, workshops and breakout groups that had to be scheduled around the all-important plenary lectures. It would take only one session to over-run slightly for the whole finely-tuned extravaganza to be thrown out of kilter. To the unsophisticated mind it would have seemed impossible that a mere 50 participants could be spread among so many events, held simultaneously, in the space of three days. But Nigel Archer well knew that idleness led to mischief and he was determined to keep the officer corps of SI fully occupied. Part A of his plan called for the working day to be extended well beyond its natural span, at both ends. He himself invariably rose at 5.00am and began his day with a 5-mile run. For those with any ambition to reach his inner circle, the early morning run was *de rigueur*. The field fell into two categories: those who 'happened' to be warming up as he emerged from the hotel and those who would artfully arrange to be running in the opposite direction. The latter was considered the safer strategy, as it not only allowed them to bid him a cheerful 'good morning' as they passed, but gave them the freedom to select the duration and pace of their own exertion. Those who chose to accompany the Chairman were expected both to match his impressive pace and respond to the questions he invariably fired at the members of the escort. Miel, it goes without saying, knew of these arrangements and had determined that she would be in the latter group. She had, accordingly, been training intensively for the last two months. By the time she arrived on Cap Ferrat, her cardiovascular system had reached such a peak of performance that she could have run the course while singing Turandot, and still had breath to spare.

The 'official' day began with a working breakfast at 7.00am, followed by sessions, both parallel and plenary until lunch (at which 'case studies' were discussed; then more sessions, mainly parallel, until 18.30, when there was an official break until dinner at 19.30, during which the Members of the Board 'table-hopped' between courses, followed by an after-dinner speaker, whose words were invariably designed to stimulate 'out-of-the-box' thinking. In theory, the proceedings ended at 22.30: but only for the suicidally unambitious. Those with any interest in their careers would be working into the wee small hours to make sure their 'homework' for the next day was completed. You never knew when The Chairman was going to 'hop' into your particular Working Group or Case Study.

Section 3: *Awards Dinner:* A warm glow crept over Klomp's heart as he read this and he demanded another gin and tonic from the stewardess to enhance the sensation. A glance down the page confirmed the Snurt's now accustomed position as the climax of the evening. Lesser Awards, such as 'Innovator of the Year', 'Best New Product', etc. were dispensed with between the soup and the main course – no speeches beyond a brief thank-you were either allowed or even possible amid the fluttering of waiters bearing dishes to and fro. Yes, the Snurt was the highlight of the event. Not until the coffee had been drunk, the liqueurs served and the cigars lit was the winner announced. And Klomp was ready for his moment of glory – a moment which he knew would last 50 minutes and involve a minimum of 25 slides. His acceptance speech was by now word perfect, even down to the 'impromptus' that would be his devastating-yet-oh-so-witty responses to the joshing-but-warmly-collegial remarks he would receive spontaneously from the audience.

Although his bosom buddies (neither had accepted his invitation to attend the dinner) would not be present, Miel had scripted some appropriately risqué remarks for selected Management Trainees to interject at appropriate points in the Proceedings.

Henri de Klompenmaker looked forward with relish to the coming days. He lived for these meetings; he loved their pointless hyperactivity and feeling part of the SI elite. He felt he always shone on these occasions; ever-ready to present; always eager to ask a question; working into the small hours on his contributions for the next day's sessions. It would be such a contrast with Bermuda. No idling by the pool here! But Bermuda had been useful in one very important way. There he had met Dr Lola Santiago, this year's Guest Speaker. Klomp, whose academic career had progressed no further than a master's degree, had been dazzled by this glamorous Nobel Prize-winning economist and was keen to bathe in her reflected glory. Perhaps he should offer to introduce her to everyone? He had already asked Lionel Grove to seat him next to her at dinner on the first evening. Lionel's abrupt 'Out of the question' might discourage some, but not Henri de Klompenmaker. He was sure that Lola, who would undoubtedly remember their brief encounter in Bermuda, would insist. As he closed his folder, the pilot announced that they would shortly be arriving at the airport of Nice Cote d'Azur and that if they cared to look out of windows on the left-hand-side, they would see Cap Ferrat below. Klomp looked down at that benign green cape, canopied in pinewoods, speckled with sun-kissed villas and fringed by a rocky shore, which hinted at a world apart. He scanned its East Coast; he scanned its West Coast; he scanned its Southernmost Tip. 'Doesn't have much of a beach', he concluded.

As the aircraft taxied to its stand, Klomp could see a large white limo on the tarmac, near where the plane would come to rest. 'Must be a VIP on board', he thought. Then he spotted the man wearing a large sun hat. It was Lionel Grove. He heard the rising chatter from the management trainees several rows in front. 'He's here to meet the Chairman! Somebody's cocked up'.

'Yes', Klomp thought, 'somebody has cocked up, and I bet its not Lionel'.

One of the management trainees persuaded the stewardess to open the door of the jetway and attract Lionel's attention. He mounted the stairs to be told that The Chairman was not on this flight'.

'I know that! I'm not here for the Chairman. I'm meeting Dr Lola Santiago. She's due to land any minute. Have I been dragged up these steps for nothing?' Nobody dared answer.

At that point, the sharp-eyed observer would have seen a small plane looping the loop a few miles offshore. Dr Lola Santiago was amusing herself until she was given permission to land. Lionel returned to the limo, opened the door and took refuge from the hot afternoon sun.

By the time the hotel shuttle disgorged Henri de Klompenmaker and his colleagues into the forecourt of Hotel Excelsior they all knew of his acquaintance with Dr Santiago and were looking forward to being introduced to her by her number one friend in publishing. The Excelsior was a grand hotel of the old school; its calm exterior belied a hive of activity within. No sooner had the bus stopped than a troupe of porters emerged from the building, surrounded the vehicle and with startling efficiency removed the luggage to the interior of the building. This brisk efficiency was to set the tone for the entire SI sojourn *sur Cap*.

The SI Corporate Strategy Meeting Delegate, while checking in, could hardly miss the magnificent seascape that opened through the enormous picture windows in the hotel lobby. But a 'Steward' would soon direct him to the small room that was the location of the 'Meeting Registration' desk, where his two-volume Full Programme awaited him, along with stern instructions that the 'Welcome Reception' began at 18.30 sharp. Having been encouraged to proceed straight to his room, he would find there an even more spectacular view: a prospect that extended from Cap Martin in the east to Cap d'Antibes in the west. Immediately below the lush hotel, gardens seemed to extend to the very edge of the sea. And was that a funicular railway in the middle distance? Yes it was, a *'funiculaire prive'*, exclusively for the use of those guests who would make use of the hotel's swimming pool, but could not summon the energy to negotiate, unaided, the modest slope that led there. Enticing, but not an option for the newly arrived SI 'Delegate', who would, if he knew what was good for him, devote the time between now and 18.30 to a close study of his Full Programme, which would reveal the plan of action for the next three days. The Chairman believed strongly in the element of surprise. This was most famously exhibited in is highly unpredictable and often nocturnal telephone habits, but was also evident in the organization of the Strategy Meeting. He liked to do this in what he called 'real time', meaning that no delegate knew which Sessions, Workshops, Encounter Groups or Case Studies he would be involved in until he opened the Full Programme. Ignored were the stupendous views, the evening air heavy with the scent of blossom, even the seaside swimming pool reached by its *funiculaire prive*. Corporate advancement, and even survival, could hinge on performance at this meeting. The Chairman liked to encourage what he called 'healthy competition' among his executives. This meant that they were expected to tear strips off one another at every opportunity. The late afternoon hours on that first day were, according to the Full Programme, for 'orientation', but only the foolhardy would use this time to explore the Hotel Excelsior's extensive grounds or facilities. Consequently, an eerie hush descended, broken only at 18.25, when a stampede of SI executives headed towards the Somerset Maugham suite for the welcome reception, which began at 18.30 sharp.

There they would find Lionel Grove at the door, surveying the dress of each delegate. Those he deemed properly attired would be directed to The Chairman. He was the room's centre of gravity, and around him they circled. Circled like sharks waiting for the scent of blood as various individuals were summoned to the Presence to be loudly praised or equally loudly harangued.

By 19.15 the Chairman was ready to go into dinner. 'Where is she?' he asked Lionel, for the sixth time. 'We're paying her enough. She should be on time'.

'I know she had a private tennis lesson when she arrived, and she was last seen heading to her suite to have a cup of tea with her instructor. But that was more than an hour ago. She's not been answering her phone'.

'Well, go up and fetch her. We're going into dinner in five minutes'.

With that, the room fell silent. All heads turned towards the door. Dr Lola Santiago had appeared.

'Lionel, darling!' She threw her arms open and stamped her feet playfully. All eyes were on her, or rather on her dress, at least what there was of it.

Lionel, delighted to be recognised by the star of the evening, hastened to greet her, receiving a hug and an air-kiss to both cheeks. They quickly surveyed each other. Both were connoisseurs of the art of cosmetic surgery.

'I should take you over to The Chairman. He's waiting for you'.

'OK darling, but I want to talk fashion later. I love your jacket. Armani?'

'Yes. *Caro Giorgio* is so generous.' First point to Lionel Grove.

He escorted Lola to meet the Chairman who, along with all the other men in the room, could not take his eyes off the contents of the guest speaker's dress.

Introductions were made, and only the briefest pleasantries exchanged before Lola took Lionel to one side.

'Nigel, darling, please excuse me one moment, I need to speak with Lionel about the arrangements for my talk after dinner. Lionel. You got all my slides, yes?'

'Yes, and they are all in the projector'.

'Good. Who's operating the projector'.

'One of our stewards. He knows what he's doing'.

'Good. Now, I have a few unscripted things I want to say before the slide show –OK?'

'Unscripted?' Lionel Grove did not like the sound of that. 'I don't know. You should clear that with The Chairman. He's approved your presentation already and he doesn't like surprises'.

'Hey! I guess he's paying Dr Lola Santiago to say what she thinks. I give him my thoughts'.

'Well, I don't know. I think you should check with him first'.

'Lionel darling, these are things you guys at SI need to know. No need to check. Let's go to dinner. You like my outfit? First time on'.

Lionel did not approve of middle-aged women showing quite so much flesh.

'Stunning. Versace?.'

'Of course, darling. I advise him on pricing strategy. He pay me in clothes. Too bad SI got nothing I want! You guys gotta fork out the cash'.

The dinner was especially enjoyed by all those at the Chairman's table, where Lola Santiago was at her most entertaining. Her life story was exciting and needed no embellishment. After escaping on a home-made canoe from Castro's Cuba, she won a scholarship to Harvard and spent a year as an intern in the Kennedy White House.

'He wasn't such a great lover. Always in a hurry'.

None of them had an adequate response to this or any of her other anecdotes about Kissinger, De Gaulle or Maxwell. By the end of the dinner, however, there was a silent consensus that she had earned her enormous fee already.

'You should write your memoirs', urged The Chairman.

'Only when I've sold the movie rights. Spielberg's very interested, but he's too cheap. I tell him "Don't call me back till you're ready to talk in seven figure sums. This is Lola Santiago, not some freaky extraterrestrial!"

At the end of dinner, following an effusive introduction from The Chairman, Dr Lola Santiago rose to speak.

'Mr Archer very kindly invites me to tell you about pricing strategy and I'll get round to that. But let me tell you guys that before you can talk about prices, you gotta get your product right. I tell you here today that you got a long way to go'.

At this point, the Chairman rose to say 'Dr Santiago, I think it would be better if you stuck to the subject of your presentation ...'. But he was cut short by the repeated stamping of his guest speaker's feet. Lionel Grove groaned and looked away. The Chairman, who had never before been subject to a public foot stamping, sat down. Dr Santiago continued.

'Let me tell you guys. You got a big problem. I mean a big problem. I don't know anything about defence equipment, elevators or shipbuilding, but if the stuff you make in these divisions is as bad as your publications, I tell you I don't feel too safe'.

'I, Lola Santiago, publish a lot, but never in SI journals. You know why? Because they're crap'.

This assertion prompted gales of laughter throughout the room. The other SI divisions were sick of having the publishing division's profitability rammed down their throat at every corporate meeting. Their representatives felt they could safely revel in this moment of unaccustomed candour in the safety of the darkened room.

The Chairman tried to speak again.

'Hey, Mr Chairman, pipe down'. Lola again stamped her feet several times. 'This is Lola talking to you. Listen and learn. I'll be brief. Your journals are crap because they're cheap in every way except the price. Cheap paper, cheap illustrations, cheap editors. I'm talking cheap. OK, you get plenty papers, but not from the likes of Lola Santiago. You guys get the crap, and there's plenty of that around, I can tell you. Hey, get out of the garbage recycling business and into publishing. Then, just maybe, you can justify your prices. I publish all my stuff with Albany. They treat me like a lady. That's all I gotta say, and I won't charge extra for it, but you should listen to Lola'.

Total silence prevailed. Never had anyone challenged the ethos of SI Publishing so publicly before. It was the most profitable publishing business in the world and here was she, a Nobel Prize-winning economist, telling them that their product was crap. Nobody dared speak.

The Chairman thought she had gone mad.

Lionel Grove thought it was an act to shock everybody into alertness for her main presentation.

Henri de Klompenmaker broke out in a cold sweat.

Lola asked for slide one. 'This evening I tell you about my general and specific theories of pricing and also how to apply this to your businesses to optimise shareholder and customer satisfaction'. As the first slide, filled with undecipherable equations, appeared, the room sensed that they were sailing into more placid waters. By the end of her presentation, 90 minutes later, they felt positively becalmed. The Chairman, normally very assiduous in keeping speakers within their allotted time, was reluctant to interrupt this one, in case he set off another embarrassing Latina outburst. Now he knew why she was nicknamed the Cuban Firecracker. He was equally reluctant to encourage questions at the end, and brought the proceedings to a hasty conclusion as soon as she stamped her feet to signify that she was finally finished speaking. It was late and they had a busy week ahead.

Each day began long before the 7.00am working breakfast, as delegates rose early to complete their 'homework' for that day. There followed a frenzy of activity and interaction from dawn till dusk. Reputations rose and fell. At least one delegate required medical treatment after his plans for 'The Combine Harvester of the Future' were comprehensively trashed by The Chairman in front of a baying audience. Another was so humiliated by his group leader in the course of a Case Study that the subsequent drowning of his sorrows in the bar not only eliminated him from the rest of the Strategy Meeting, but seemed set to cut short his career at SI.

Klomp suffered no such indignities. The definitive 'safe pair of hands', his contributions were diligent, exhaustively prepared, in tune with the Chairman's thinking and thoroughly unimaginative. Following Lola Santiago's frontal assault on his publishing business, he was afraid that she might join some of the conference workshops. But he had nothing to fear. Much to the irritation of Lionel Grove, Lola spent all her time receiving private tuition from Sven, the resident tennis coach. She left the day before the Awards Dinner, her name never uttered again by The Chairman. As he dressed for this, Klomp basked in Miel's assessment – and she was a reliable barometer – that he had had a 'Good Meeting'. He had been seen running every morning at the required time, had picked on the right people during question-and-answer sessions, and the case study he had 'facilitated' was the very model of ill-tempered bickering that The Chairman adored. Yes he had had a Good Meeting, but the best was yet to come.

As far as he could tell, nobody knew about the Snurt. No nods, winks or knowing glances had been cast in his direction. Even Lionel Grove, who was in charge of the whole event, had dismissed him with an ill-tempered 'Of course not' when he asked in feigned innocence if he knew who the winner was. 'The Chairman's Office takes care of that one. Apparently I'm not to be trusted. I don't see what all the fuss is about anyway. The only good thing about it is the slide show. When the lights are down we can all have a good sleep ...'. He stopped himself. He was going too far in front of Klomp, about whom his friend Marlene Pym had given the sternest of warnings. 'Not to be trusted'. He changed tack, 'Do *you* know who this year's winner is?'

For some reason, this question had sent Klomp scuttling off pretty sharpish. Lionel shook his head as he watched the retreating figure. 'What have we come to? And where did he get that dinner suit. Hired, obviously!'

'Very strange behaviour', Klomp reflected as he straightened his pre-tied bow tie. 'Sometimes I think Mr Grove is not as corporate as he might be'. But he killed the thought. He dared not take any risks with Lionel Grove. Stories of people who had got on his wrong side were legendary. Offices (he was in charge of all corporate office space) had been shrunk overnight; cars (corporate transport was one of his departments) had failed to turn up at airports; airline tickets (corporate travel reported to him) which should have been business class metamorphosed into economy. No, he was not going to think negative thoughts about Lionel Grove. Now dressed in his hired dinner suit, he admired himself in the mirror. He thought he cut an elegant figure. Jackets *were* being worn very loose this year, weren't they? He could always roll up his sleeves, just like Don Johnson on Miami Vice. And the trousers: he had seen Tom Cruise wearing trousers just like them; obviously cut on the long side and casually bunched around the ankles, just as his did. And he was sure that big bow ties were back. Yes, he would cut a dash when he rose to receive his award. Miel was taking care of his slides, all 40 of them, so he had no concerns on that score. He had also rehearsed his speech every night that week and was word perfect. Yes, Klomp was ready to face his public.

A trumpet fanfare announced the arrival of The Chairman and his guests. All stood. As they proceeded to the Top Table on a dais at the end of the room, Klomp recognised but a few of them. Lionel Grove, obviously, as he was the Master of Ceremonies for the evening. The others, he gathered from Miel, were either local dignitaries or non-executive directors who had been invited to attend the Meeting.

'But I did not see any of them at our sessions'.

'Wrong place. You should have looked by the pool during the day or in the bar late at night. They were all hanging round Lola Santiago. The Chairman has to pamper these guys. They brought him in and he relies on their support'.

The banquet commenced. The first course, *Consommé a la Sevigne*, was barely finished when Lionel Grove stood to announce the first award, for 'New Product of the Year'. Amid a ripple of applause, the lucky recipient picked his way through shoals of darting waiters to the Top Table, where The Chairman leaned across, shook his right hand, thrust a package into the left and threw a smile to the cameraman from *SI Universe* who was in attendance to record the event. Within 15 seconds of his arrival, the winner was on his way back to his table, this time in silence, as the assembled hosts were already tucking into *Filets de Sole Imperiale*. At the first sign of a knife being put down at the Top Table, the waiters pounced to clear away this course, as the award for 'Innovator of the Year' was called. This recipient, a slight blonde, was all but swept into the kitchen by the wave of waiters that she had to fight to reach the Top Table. She had her 15 seconds of fame to the sound of silver covers being lifted to reveal *Noisettes d'Agneau Chatelaine*. Between the *Noisettes* and the *Cailles sous la Cendre* which followed, the time taken for the necessary change in accompanying vegetables from *Petits Pois* and *Pommes Nouvelles* to *Salade de Laitues* and *Asperges Verts* gave Lionel a welcome opportunity to fit in two awards. Nobody could fault his logistics. Nine awards had to be fitted into a dinner of seven courses, so he doubled up where he could. The unfortunate collision between the arriving recipient of the 'Marketeer of the Year' and the departing recipient of 'Salesman of the Year' was due less to the traditional rivalry between these two departments and more to the dazzling effect of flashlights dancing on dozens of silver covers being carried at head height. They were quickly helped back to their feet, by which time the attention of the assembly was given over to the dismemberment and consumption of tiny quail corpses. This completed, Lionel cleverly took advantage of the rearrangement of cutlery necessary for the *Mousse Glace aux Fraises*, which was next on the menu, to fit in 'Innovator of the Year' and 'Best Marketing Leaflet'. These recipients, more agile than the previous 'double' managed to regain their seats without injury.

At this point, Lionel himself was distracted by a waiter with a note for The Chairman, who looked at it, frowned and shook his head. The waiter retreated and Lionel prepared for the next award, for Management Trainee of the Year, which was to follow the mousse. Miel beamed, recalling her own triumph last year, when she had won this award. Klomp began to perspire even more heavily than usual. Only one more award before The Snurt – 'Office of the Year'. That was a surprise; Klomp thought that the Defence Systems office in Australia was to be closed to save costs. The working group he sat on had recommended it; the money saved on travel alone would be enormous. Unusual for The Chairman not to act on that type of advice. He could never resist a chance to cut costs. The Chairman too, looked surprised – annoyed even – as he curtly passed out this award. (The Australian Office was closed six months later).

The Coffee was served, liqueurs poured, cigars lit. The Chairman rose once more, this time to utter his first public words of the evening. 'Dear Colleagues', he began'. Rewarding excellent performance … a first class company…

another challenging year … it's all about delivery … people are our greatest asset …you can't break an omelette without breaking legs, I mean eggs … we turn now to our most important award …for administrative efficiency … this is what makes our profits … the winner of this year's Snurt Award is … … Henri de Klompenmaker.

A stunned silence greeted this announcement. The Publishers were stunned; they did not even know that the Publishing Division was eligible. Elevators were stunned; Klomp, who had been one of their own had defected to publishing and they knew that he was now enjoying to the full, the high spending publisher lifestyle that they deplored. Lionel was stunned. He instantly recalled that conversation the day before, when he had let slip to Klomp his real views on the Snurt. Had he said anything too damaging? Quickly rifling through the well-organized filing cabinet that was his mind, he concluded with relief that he had not gone too far. Anything that Klomp could transmit to The Chairman was relatively innocuous and could be denied.

'Lionel, Lionel!' It was the Chairman, recalling him to the present and wanting assurance that the projector was ready. It was, he assured him. Henri de Klompenmaker approached the podium, decked in the Snurt medal and clutching the text of his Acceptance Speech. The lights dimmed – 'Figures are everything', Klomp began, as the first slide, a three-dimensional multicoloured graph, appeared on the screen. By the fifth slide – 'There is nothing that cannot be measured quantitatively'. The sound of bottles being circulated and glasses drained, could be heard throughout the room. Lionel, who could see nothing in the gloom, congratulated himself on his foresight; he had made sure that every table would be well stocked with bottles of brandy and port before the Acceptance Speech began.

This would keep the troops from becoming too restless. By slide twenty-five – 'A new tool for measuring journal performance', Lionel was sure that he could hear snoring from the back of the room. Striving to keep awake himself, he was startled by a waiter trying to push past.

'What is it?' he hissed.

'Very urgent message for Mr Archer'.

'He can't be interrupted now. He already sent you away'.

'Very urgent. New York on phone again. Insist they must speak to him'.

'Let me talk to them first'. He welcomed this interruption, not only to escape from the tedium of Klomp's speech, but also to find out what was going on. Lionel liked to keep abreast of corporate affairs. Also, this waiter was quite cute.

He pushed the waiter towards the door, crossed the lobby and informed the reception desk that he would take the call.

'Hello, Lionel Grove here'.

'Where's Archer?'

'He is busy and can't be interrupted'.

' Buddy, I don't know who you are, but I've got to speak to him now. We need to get a message out to the markets'.

Lionel was not 'buddy' to anyone except, well, certainly not to this person. 'Tell me what you want and I'll speak to him. I'm a director, you can tell me'.

'OK, this is Bas Lehman from Filstein Brothers. There's a rumour circulating in the US about SI. The stock price is falling in New York. We need a statement to reassure the markets. And we need it now'.

The words 'stock price' hit Lionel right between the eyes. He had a lot riding on his SI stock options. Most notably, that villa in Mykonos that he was planning to build for his retirement.

'I'll get him straight away. Hold on'.

'He hastened back to the Somerset Maugham suite. It was but a moment's work to catch The Chairman's ear and persuade him of the urgency of the situation. He had even more riding on his stock options than did Lionel and he easily outpaced him as he headed for the phone. They weaved through the tables – a voice in the darkness asked 'Is he coming back?'

'I think not!' hissed Lionel in the direction of the unknown voice. He knew that the words 'stock price' hit the most sensitive of corporate buttons; when combined with 'falling' he knew that a corporate crisis was in full swing. The Chairman would need his support to deal with it and by the time he reached the door, Lionel was in full crisis management mode. Tonight's events would make a very interesting diary entry.

Henri de Klompenmaker plodded on through his speech. Entirely absorbed in his own, word-perfect delivery, he was unaware of the dramatic events that were unfolding around him. The glare of the projector meant he could see nothing, but from about slide thirty – 'More sophisticated numerical tools', he became aware of a constant background noise of chairs being moved and the fall of footsteps on the parquet floor. More alarming was the absence of the choreographed 'interruptions' that had so enlivened the first part of his speech. Though the interruptions did not come, one or two of his 'impromptu' responses to them did; the resulting titters reassured him that his wit remained razor sharp. He missed The Chairman's signature loud guffaw, but nothing was going to throw him now; he was on the home straight, in full flow and word perfect. The final slide appeared, accompanied by his last, casually thrown, witticism. The lights went up. He faced a room empty, except for three or four somnolent diners and Miel, who applauded enthusiastically. Klomp was stunned. He looked to the Top Table: no Chairman, no Non-Executive Directors, no Lionel Grove. He surveyed the room. Where had they all gone?

Miel beckoned him to her table, where she breathlessly explained everything.

The Chairman had been summoned to the phone to deal with a corporate crisis. The sound of feet on parquet that had been the accompaniment to the second half of his presentation, initiated by the exit of Lionel and The Chairman, had been taken up by the Non-Executive Directors, whom The Chairman had commanded to his presence, followed by a steady stream of other participants who felt that they should be on hand in case they were needed. Miel's mention of the words 'stock falling in New York' were sufficient to persuade Klomp of the severity of the crisis and her soothing assessment of his speech, 'The Chairman seemed to enjoy the first half', was enough to convince him that it had been a great success.

Klomp was by now exhausted. He could take no more excitement and headed across the lobby to the elevators. Yes, he would have an early night tonight. No sessions to prepare for and no speeches to rehearse; he would at last have time to read the latest issue of *Stamp Monthly*. As he waited for the elevator, he could see his audience in the bar, on hand for the Chairman. Miel bade him a good night and went to join them.

The talk next morning, the day of departure, was all of the crisis. The Chairman had been on the phone all night and had taken the early morning flight to London, where the SI stock was now falling. The markets had found out that the Natural Resources Division had been systematically overstating the its profits for years. The executives from that division were also on their way back to London in The Chairman's entourage. Their ritual execution was inevitable, but they received few expressions of sympathy from colleagues. The circling sharks were waiting for the opportunities that would arise from this shipwreck.

Henri de Klompenmaker knew what it meant for him. 'They'll want more profit from us to fill the gap. I'll need to cut costs. If I deliver, my next promotion should be a good one'. Klomp already had a position in mind that he knew would suit him perfectly. During his morning run the previous day, he had heard that the Presidency of the SI Leisurewear Division was shortly to become vacant. Klomp was not one of nature's athletes, but he would be able to buy the leisure suits of which he was so fond, at a reduced price. Then there were the complimentary tickets to major fixtures and the prospect of rubbing shoulders with the sporting glitterati of the Low Countries, who sponsored the products. That would impress his drinking companions in the little local bar he missed so much. Publishing was not a regular topic of conversation there, but sport was, and now they would listen to him. He'd move back home. Motivated by the prospect of being a big shot in his little town, he began work on his cost-cutting measures. Personnel, the costliest item of all, came first. Klomp had always kept a mental list of staff who could, when required, be sacrificed for the advancement of his career. Now was the time to refresh that list. The first name required little thought: 'Fiona Hamilton'.

Chapter 10: Fiona Visits the Coast

'Fiona, Henri wants you to take a trip on his behalf. He's too busy', was how his secretary broke the news.

'Where to? East Berlin? Ulan Bator?' Her boss was in the habit of keeping the more glamorous locations for himself and the memory of her recent, high cholesterol, vitamin-free trip to Tallinn was still very fresh.

'No, California'.

'You're joking! Henri wants me to go to California in his stead? Where in California?' Fresno? Bakersfield?' Fiona was deeply suspicious.

'Bear'

'Where?'

'Bear! Bear State University. Henri has been invited to attend the Board Meeting of the Directors of the Todd Institute for Windsurfing and Moleculetics'.

Fiona's heart skipped several beats.

'Are you sure?' She could not believe her luck.

'Quite sure. I know it's out of character, but for some reason he's decided not to go to this one'.

'But it's exactly the kind of destination he always keeps

to himself'.

'I know. It doesn't fit, does it?'

Fiona recognised a fellow-Klompenologist. 'Too right it doesn't fit. He'd never delegate a trip like this to me. Is he ill?'

His secretary laughed. 'No, but there's definitely something going on. He's deleted virtually all his trips between now and May'.

'Very out of character, I agree. Any theories?'

'Well, the Corporate Strategy Meeting is at the end of May'.

'Yes, but he says he goes to that every year and he's never cancelled meetings like this before, has he?'

'Well, I couldn't really say. I've only been working for him for six months'.

'Yes, of course'.

'But it so happens I did check his diaries for the last two years – I believe he has been to the last two Corporate Strategy Meetings'.

'As he seldom forgets to remind us!'

'Quite. Well, as far as I can tell this is the first time he has eliminated appointments wholesale beforehand'.

'Something's going on'.

'It certainly is. Miel Flick has been spending more time with him than ever. She's in there every morning. I can hardly get him to go through his mail'.

'Well, if Miel's involved, it must be something very Corporate. Otherwise she wouldn't bother. You know what Miel's like'.

'You've noticed?'

'It's hard not to. Of course, there is another possibility. You don't think they're…?'

They both laughed and simultaneously screamed, 'No!'

and agreed that even Miel Flick was not that ambitious.

'No, I'd put my money on something Corporate!' asserted Fiona.

'My hunch is that you're right, but I've no evidence'.

'Don't despair. We have a few weeks to observe and collect data!'

'What fun. I could do with a bit of excitement up here. It's so dull compared with my last job'.

'Let's spice it up then. We'll gather evidence and hold a seminar over lunch on Thursdays'.

'Great idea!'.

'So, about Bear. When do I go?'

'May. They've sent us an information pack. It's very comprehensive. Actually, it looks like a holiday brochure. Full of pictures of surf, sand and swaying palms. It *is* a university, isn't it?'

'Of sorts. Tell me, will Henri cover my expenses? I haven't budgeted for this trip and I have no funds to spare, especially since he cut our travel budgets by 20%'.

'No need to worry. Bear cover the travel costs'.

'That's the best news I've had all week'.

'Delighted to oblige. I'll pop this stuff in the internal mail and let Bear know that you'll be coming instead of Henri'.

'Much obliged. Lunch next Thursday?'

'It's a date!'

Fiona's head spun as she put down the phone, her brain and heart both in overdrive. She'd see Tom again, and so soon! What *is* Klomp up to? Something big must be going on. And what about *Transactions*? She'd get a first hand look at this Todd Institute place, the source of these short communications that were filling *Transactions*.

The Bear State University Information Pack lived up to

its billing. It was glossier than Harrods Christmas Catalogue. From it, Fiona learned that Bear provided a unique ambience for the student who wished to have a 'rounded' education, and that the quest for academic excellence was merely one aspect of a balanced curriculum that strived to equip today's young men and women for the challenges they would face in life. From what Tom had told her, the main challenge Bear graduates faced in life was how to spend the income from their trust funds. She could now see how Bear would equip them to do so. If the curriculum in California was insufficiently stimulating, the overseas campuses in the Caribbean, Italy and Africa would undoubtedly make up for it.

The next afternoon, Fiona Hamilton was working quietly at her desk when the phone rang.

'Good afternoon. Publishing Department'.

' May I speak with Dr Hamilton?' asked the American woman

'Speaking'.

'Is this Dr Hamilton?' asked the American woman, clearly surprised.

'The last time I looked it was. Let me check again. Yes, it's me'.

'I'm so sorry. It's just, I was expecting a …'.

'A man?'

'No, an older person. You sound so young'.

'Very sorry to disappoint you. How can I help?'

'I'm Monica Parsons, Personal Assistant to the President of Bear State University. I understand that you will be taking Mr de Klomp-en-mak-er's place at our forthcoming Todd Institute Board Meeting'.

'That's correct'.

'I want to tell you how much we are looking forward to

your visit with us. Will this be your first time at Bear?'

'Yes'.

'It's a very special place'.

'I know, I've seen the brochures'.

'Good. I can guarantee you won't be disappointed. I am in personal charge of the arrangements for the Board Meeting and I want to make sure that you have a good time'.

'That's very sweet of you'.

'Of course I'm sweet. I'm from the South, sugar. We need to talk transportation and accommodation'.

'I'm afraid I haven't done anything about either yet'.

'Good. Don't you worry your little ol' head about that kind of stuff. That's my job and I want to tell you what we're going to do about it. The Board meeting is on the 7th and 8th. Can you leave England on the 3rd?'

'The 3rd? Why so early? I don't know if my boss will allow that. We're so short-staffed'.

'Honey. That's why Monica's here. We have a few good ol' boys coming over from England and they need a few days to get over the jet lag. We call these few days 'Brainstorming' and we arrange a very full programme. Golf, tennis, sailing, croquet, whatever you want. I think it's essential you attend'.

'I'm coming round to your way of thinking'.

'Most people do, honey. So I take it you can depart on the 3rd'.

'How could I say no?'

'Good. That's dandy. I take it you'll be able to stay for the post-meeting retreat? We consider it an important follow-up'. It's held off-campus and this year will be at the Bear Winery and Spa in the Napa Valley'.

'Just what you need after a heavy meeting'.

'Precisely. The retreat finishes on the 12th.

'I so hate to rush things, don't you?'

'We're cut from the same cloth, honey. Now, about the flight. We have a slot to depart from Birmingham on the morning of the 3rd. Would that be convenient for you?'

'But I'm in London. There are lots of direct flights to California from Heathrow. I can catch one of them'.

'Sugar, we are talking different languages here, Your talking scheduled and I'm talking private. Y'all are coming over on the Bear corporate jet. Geography's not my strong point, but I'm told that Birmingham is somewheres in the middle of England. We need somewheres in the middle 'cos we've got people coming from all over'.

'It's definitely in the middle'.

'And you're in London? I love London. The President and I will be there next month. The Bear UK Alumni Association has its annual meeting at Wimbledon. I don't know Birmingham, though'.

'You're not missing much'.

'So I'm told. Honey, we use a limo service in London. If you can give me your address, I'll arrange for a driver to take you to the plane'.

Fiona proceeded to give her new-found friend more details than she was in the habit of providing a complete stranger.

'Passport number?'

'So we can clear you through immigration'.

Fiona provided this and details of her food preferences, allergies and bathrobe size. The hour-long discussion covered everything.

'Monica, is there anything you don't take care of?'

'Nope. Welcome to the Wonderful World of Bear'.

'I'm looking forward to it'.

'If you need anything else, just you call Monica. OK?

'I certainly will. Thanks for your help'.

'You're welcome, I'm sure. Before I go, as you're in London, there is a little something you might be able to help me with. It's a little thing I want to take care of before we come over for Wimbledon. My momma has pug dogs, and she likes to collect art, you know; pictures, porcelain, and that kind of stuff, about pugs. Ugly little things, I'd say, but beauty's in the eye of the beholder, I suppose. Anyhow, a friend told me about this neat shop in London that sells dog antiques'.

'You must mean Paws Through The Past. It's in Walton Street. My grandmother has Pomeranians, and I get her birthday presents there'.

'Do they ship?'

'Oh yes, all over the world. Martita Cecil, the owner, is wonderful. If she doesn't have something you want, she'll move heaven and earth to find it. She even has a roster of artists that will paint or sculpt your pet in any medium you want. I'll fax you the details'.

'Honey, I can see we're going to have a lot of correspondence'.

'I'm glad to be of help'.

' So long'.

'And happy trails to you'.

Fiona sat and stared at her desk for some time after this conversation. Normally one of the most animated of SI personnel, she now looked as if she had had a stroke. But her brain was in overdrive, not paralysed. 'Corporate jet', 'Sailing', 'Wineries'. Bear seemed like paradise. Too good to be true. Which it was, as she was to find out in a phone call two days later.

'Hello, I say, hello, is that Fiona Hamilton? I want to

speak to Dr Hamilton. Vera, I'm sure this isn't the right number'.

'Les, this is Fiona Hamilton'.

'Fiona, is that you? Good. I have been trying to speak to your Mr de Klompenmaker for several days now, but apparently he is much too busy for the likes of me. Any road, his secretary tells me you'll be taking his place at the Todd Institute Board Meeting. I have to say that this is a very rum do. A very rum do. I pulled a lot of strings to get him invited and they won't like him dropping out just like that. I say, they won't like it at all. Not at all. They took a lot of persuading to let in a businessman in the first place. It's a very academic meeting. *Bona fide* scholars only'.

Fiona was holding the phone at some distance from her head while this rant exhausted itself, so he would not have heard her involuntary shriek at this last comment. 'Les Fyfe' and '*bona fide* scholar' were not terms which sat comfortably together.

'I feel very let down and I know the Todd will be most upset'.

'They seemed fine about it when I spoke to them earlier this week. In fact, they couldn't have been more helpful'

That stopped him in his tracks.

'You've spoken to them yourself, then? This is highly irregular, I say. Highly irregular. Vera and I usually co-ordinate all the arrangements for the British participants. What's gone wrong this time?'

'Arrangements seem to be proceeding smoothly, as far as I can tell. Monica Parsons told me that …'

'You've been in touch with Monica Parsons herself. I say, you shouldn't have done that. You shouldn't go bothering Monica Parsons. She's very busy, she's the Bear President's PA, you know. Even I can hardly get through to her when I call'.

'She called me'.

Silence.

'Les, are you still there?'

Les was till there, but had been rendered momentarily speechless by this news. Monica Parsons had never called him (and she never took his calls). He had to make do with the Dean's secretary.

'Les?'

'Yes, I'm still here. Listen now, Vera received a fax from Dean Huntingdon's secretary that gives all the travel arrangements. I'll have it sent to you. The Bear plane departs Birmingham Airport at 11.30 on the…'.

'I know all that. I got the same fax … from Monica'.

Another silence. Fiona allowed her last statement to sink in.

Les martialled his forces for a response.

'You seem very chummy with Ms Parsons! I suppose she told you who else would be on the flight'.

'Actually, she didn't. All she said was that there would be ten passengers'.

'Well, apart from myself and Vera, Ricky Evans from Peckham Poly, Dave Butcher from Cowley University, Norrie Dick from Jarrow College, and Cliff Matthews from Cleethorpes Tech will be there. We'll have a grand time'.

'Are they all on the Board of the Todd?' Fiona could hardly believe her ears. These were four of the legendary mediocrities of British moleculetics, whose intellectual lights shone but dimly in the backwoods of the academic world.

'Of course they are. I had them appointed. Did you say ten passengers? Who can the others be?'

Les considered himself the chief plenipotentiary of the

Todd in the United Kingdom, and took a highly proprietorial interest in the international relations between the two. He preferred all contact to go through the proper diplomatic channels, that is, himself. This was one gravy train of which he was to be in sole charge and he resented the presence of passengers not personally vetted by him.

'We'll be seeing you on the 3rd, then. It looks like you won't be needing any help from me. Now, can you put me through to Lionel Grove's office. There's something I need to discuss with him'.

'I think he's away at the moment. Cap Ferrat. Making arrangements for next month's Corporate Strategy Meeting. Can I help you?' She knew that she couldn't, but had fun asking. Les would be wanting to discuss matters of high finance, probably involving Switzerland, with Lionel.

'Oh no, no, nothing to do with you. Let him know I phoned'. He quickly hung up.

'Vera, I say Vera, you'll never believe this. It seems there are to be ten passengers on the Bear flight'.

'Ten, but we've only got seven on the list'.

'Fiona says ten'.

'How would she know?'

'You won't believe this, but Monica Parsons told her'.

'Monica Parsons! When did she start being so helpful? I can hardly get a civil word out of her. Remember how awkward she was last year over the Bear Guest House wine cellar'.

'I reckon she'll have us all frisked this time before we step back on the plane'.

'We'll be all right as long as we pack it well. I'll have a word with Norrie Dick; tell him to bring a bigger trunk this time. If he hadn't been so careless we'd have never been found out'.

'Well, to my way of thinking, we deserve every drop. Bear should be grateful for the prestige the Todd brings to them; all thanks to you, Les'.

'Never a truer word was said. But we'll have to watch that Fiona Hamilton. She's beginning to get very nosy about the Todd research work we publish in the journal. I don't like it. She could cause a lot of mischief out there. I'm not happy. Not happy at all'.

Fiona, at this point, was no happier. 'Nine hours on a plane full of free booze with Les Fyfe and his drinking pals! I had no idea they were all involved in the Todd'. The lure of this trip was beginning to fade. 'And a week or more in California with them. The things I do for SI'. As she turned over the prospect in her mind it remained very gloomy, and yet through the clouds, the glimmer of an opportunity began to emerge, to find out more about what was going on. Les was, she knew, a deeply unpleasant drunk, but he was also an indiscrete one, so she could pick up some useful information. Too bad Vera would be there too. She would be on guard duty. Still, nine hours is a long time. Even Vera Fyfe would doze from time to time. And Les would be surrounded by his disciples, which, she had observed, when combined with drink, tended to bring out his boastfulness. Yes, if she kept her head, this flight could be very informative.

One misty, grey English May morning, the chauffeur driven Daimler swept up to the bottom of the aircraft steps. A uniformed steward awaited its arrival. As Fiona Hamilton stepped onto the tarmac, the trunk opened and the steward reached in for her baggage. He rummaged deeply, but emerged with only one case.

'This all you've got, Miss?'

'Fraid, so. I travel light'.

'A welcome change, if I may say so, Miss. I'm Paul, and I'll be looking after you on this flight. Please step this way'.

Fiona followed him up the steps and into the plane. It was much larger than she expected. She had seen corporate jets before, but nothing like this.

'I thought we were going to be more cramped. This is like an airliner'.

'This is the Cadillac of corporate jets. It can take 25 passengers in supreme comfort. You're going to have a great flight'.

'When Fiona stepped though the door, she felt like she had walked into a crème caramel. Everything was soft and creamy. The ceiling, walls and floors were all covered in a pale, cream carpet. The tables and trimmings appeared to be a cream marble and the seats, well, they were upholstered in the plushest, softest, creamiest leather she had ever seen. She noticed that Paul's uniform matched exactly the colour of the interior trim, and was a bit alarmed when part of that trim seemed to move towards her.

'This is Hank', said Paul, introducing the other steward.

'Good morning Ma'am. Would you like to freshen up?'

'Yes, please'. She didn't really, as the 90-minute drive from London had been one of stress-free, air-conditioned comfort, amply supplied with coffee and that morning's newspapers. But she thought it was the right thing to say.

Fiona lingered in the bathroom, unable to decide between the various designer colognes on offer. After some experimentation on various pulse points, Bulgari won out, and she emerged, fragrantly, into the cabin.

Paul walked her down the aisle. 'Three of the other

guests are already here and we expect the others any minute.'
Fiona braced herself, but was relieved not to recognise any
of the youthful faces that looked in her direction as she
approached.

'May I introduce Dr Fiona Hamilton, from the Standard
International Corporation. Dr Hamilton, may I introduce
Miss Cordelia Young from Oxford, Miss Henrietta Wolfe
from St Andrews and Mr Hugo Thynne from Exeter. They
are this year's Todd scholars and we're delighted to have
them on board. Dr Hamilton, as the plane is less than half
full you can have a choice of seating. Professor Fyfe and his
colleagues, whom I understand you know well, like to sit on
the couches at the other end, near the bar, and there's plenty
room down there for you'.

'Oh no, don't worry about that. Is this seat free?' She
pointed to a particularly well-upholstered armchair next to
Miss Henrietta Wolfe.

'Of course, darling. Do join us, and call me Hen'.

' And I'm Winky', said Cordelia.

'And I'm Gussie', said Hugo, completing the second
round of introductions.

'And I'm Fiona, as you might have guessed'.

'Would you care for refreshment, Dr Hamilton?' asked
Paul, restoring some formality to the proceedings.

'Oh, yes, a cup of tea, please'.

'Nothing stronger?'

'Not yet, thank you'.

'Isn't this something? Said Hen.

'We've been scanning the wine list', added Winky.

'And have decided to have Krug as an aperitif, followed
by a Chateau Lafite Rothschild and a Bear Winery Muscadet,
in honour of our sponsors', added Gussie.

'I think I can go along with that. But I'll have my cup

of tea first'.

'What takes you to the Todd?'

'Board Meeting. I'm attending on behalf of my boss, who can't make it'.

'Nice one. Have you been to Bear before?'

'No, but I've heard a lot about it'.

'Me too. I still can't believe I won a scholarship. It's the top facility in the world. The competition was so stiff. Over two hundred British applicants for just three places'.

'Winky, you're one of the best. You were always going to get it'.

'Thanks, Gussie. You're so sweet'.

Fiona could hardly believe her ears. They looked like intelligent young people. Competing for scholarships to The Todd? The top facility in the world? Poor deluded creatures: could they be that naive? Would it be kinder to tell them now, while they still had time to disembark? No, they'd discover the awful truth soon enough: in fact, as soon as Les and his pals boarded. Fiona decided on small talk.

'And what will you be working on when you are there?'

'Well, I need to work on my jibe technique'.

'And I need more rough water practice'.

'I'm trying to build up for the endurance events'.

'What? I don't quite understand'.

'Windsurfing. We're all Franklin Todd scholars'.

'Oh, I thought you were scientists going to work in the Todd labs. What a relief'.

'That's a laugh. Hen, in a lab? Her manicurist wouldn't stand for it'.

'Actually', a rather indignant Hen replied. 'Actually, my problem is with my hairdresser; all that salt water is so drying. But it's worth it for Queen and Country. Windsurfing is tipped to go Olympic and we all want to be in the British team. The Todd is the Mecca and has facilities that nowhere else can come close to. It will be a nice change from the North Sea'.

'And from the English Channel'.

'Not to mention Farmoor Reservoir'.

Fiona now recalled that Tom had mentioned the Todd windsurfing facility. She had read about it in the Bear brochure, but any enthusiasm she might have had for sports had been knocked out of her, literally, on the hockey fields of her boarding school. Her indifference changed to amazement as Gussie, Hen and Winky regaled her of its wonders. None of them had actually been there, but in the world of windsurfing, the Todd had precisely the opposite reputation to the one it had in science.

Winky eventually changed the subject. 'Who is this Professor Fyfe? I've heard the stewards talking about him. Not very nicely either'.

'People have been trying to find something nice to say about Les Fyfe for years, but I don't think anyone has succeeded'. Fiona, over tea, provided brief biographies of the passengers that were shortly due to join them. Gussie entertained them with an impression of Vera Fyfe's walk, as described by Fiona. Their laughter was cut short by the sound of a car drawing up outside and they rushed to the windows.

'No, that's not him. That's Ricky Evans'.

'Peckham Poly?'

'Correct. Oh and that's Dave Butcher with him'.

'Cowley University?'

'Yes, they must have shared a car'.

'Is that chap going to make it up the steps? He doesn't look too healthy'.

'Look, Hank's pushing up him from behind'.

Fiona watched as Professor Butcher made his slow progress up the steps. He looked even fatter than when she saw him two months ago in Estonia. They could hear his panting as he entered the cabin, but taking Fiona's advice, all three scholars kept their heads down. The much fleeter Dr Evans had already made his way to the bar and had set up brandies for both of them.

'May I introduce you to our other guests today?' enquired Paul. 'Some arc already here'.

'All in good time, all in good time. First things first. What's on the menu today?'

'Would you care to survey it now?'

'Bring it! And the wine list. I'll probably be wanting a massage as well. Tell Jade, won't you'.

'I'm afraid Jade isn't with us this trip. Just Hank and me'.

'I'd say that's a major oversight. Well then, let's have a look at his menu'. Ricky Evans knew the form on the Bear jet. Not surprising, as this was his third journey on it. Once Professor Butcher had regained his breath, the two of them settled into a lengthy discussion of the wine list. At the far end of the cabin looks and nods were exchanged, but no words.

Another car drew up. It disgorged Cliff Mathews, who joined the others at the bar. Eventually, at 11.25am, a white, stretch limousine approached the plane. The Fyfes were asserting their privilege to arrive last. At the front of the cabin, out of sight of the bar, four faces were pressed to the windows.

'Look at these trunks. These people don't believe in travelling light, do they?' Gussie made a mental note of Vera's waddle as she mounted the steps, her husband keeping a safe distance behind her, lest gravity reassert her rights.

'Oh look, it's Norrie Dick', whispered Fiona, as a third body, much thinner than the other two, emerged from the limo.

'He looks like a weasel'.

'He is'.

Les and Vera entered the cabin to much halloing, back slapping and the clinking of glasses.

'We set them up for you'.

'You're a mate, Ricky. The brandy in that limo was not up to much I can tell you. Are we ready to go? I say Paul, are we ready to go?'

'As soon as we get the trunks on board. It'll take a few minutes. They're very bulky'.

'I don't see Fiona Hamilton. Well, if she can't be on time, we'll have to leave without her'.

Fiona thought that she should declare her presence.

'Hello Les'. She stood up and waved to him from the far end of the cabin.

'Oh, hello Fiona. You made it, then'.

'I got here 45 minutes ago. Would you like to meet our fellow passengers?'

The other three stood up and made their way to the bar. Les and his cronies seemed very pleased to meet at least two of them and were quick to offer Winky and Hen drinks. Vera was evidently less impressed.

'Come on now, Les, we should all get into our seats. We'll be taking off shortly.'

'See you later, ladies'.

The first two hours of the flight passed agreeably for Fiona Hamilton. She and her three companions made a jolly group, a safe distance from the Fyfes. The two parties drank their aperitifs and ate their separate luncheons. This was useful, as well as pleasant, for Fiona, as it gave her an opportunity to discuss the Todd with her travelling companions.

'So, most of the Todd pyramid is actually laboratories, then?' asked Hen, looking at the glossy photograph in her brochure. 'I'd no idea'.

'Nor has most of the *bona fide* scientific world', Fiona interjected.

'I hope it doesn't smell'. Winky was very sensitive to smell. 'I'm going to California for lots of fresh air and sunshine'.

'That shouldn't be a problem. Rats are actually very hygienic'.

'Rats!'

'Laboratory rats, on which they test stuff'.

'Poor things. For the good of mankind, I hope'.

'That's what they'd have us believe. They claim that their stuff helps you lose weight'.

'That'll benefit womankind, anyway. Just what we need. When can I get it?' Winky was a full-figured girl.

'Well, it's a bit more complicated than that. And Les Fyfe is more involved in this than he's letting on. I just know, and I'm going to get to the bottom of it'.

'How wonderful!' shrieked Hen. 'I love a mystery'.

'Do you think, if I'm nice to him, he'll get me some of this magic potion?'

'No problem, Wink, didn't you see the way he was leering at you and Hen earlier on?'

'Gussie, you're just jealous'.

'Of that blob, hardly. Anyway, I'm saving myself for all the Californian babes'.

'As if they'll be interested!'

'At least I don't freckle'.

'That was unkind, Gussie. Hen's got a delicate English complexion'.

The three Todd scholars entered into a lively discussion on the pros and cons of a fair skin. Fiona interrupted the debate.

'Actually, you've given me an idea for some Miss Marpleish fun'.

'Do tell'. Hen was eager for some in-flight entertainment.

'Well, Hen, as Gussie has pointed out, Les could not take his eyes off you and Winky'.

'They were popping out of his head'.

'Precisely. My idea is (tell me if you don't like it) that you and Winky could join Les and friends for a drink at the bar, once they've had a few, and then you can, shall we say, ingratiate yourselves with him'.

'For the good of science, of course'.

'What fun. I'll ask him for some magic potion'.

'More subtle, please! If you blurt that out, you'll put him on his guard'.

'Fiona's right, Wink, we've got to be crafty'.

'What are we meant to be finding out anyway?'

Over dessert, a *Soufflé au Grand Marnier*, Fiona told them. Luncheon finished, they became aware of a general movement at the other end of the cabin. Rest room doors opened and closed. Vera announced that she was going to have a 'lie-down' and waddled back towards the sleeping cabin. Wan smiles and nods were exchanged between her and Fiona.

Vera's departure was a signal for Les and his cronies to gather round the bar.

'Paul, mine's a brandy, best you've got' was the signal

the girls to make their move. Fiona winked at Wink, who winked at Hen. They both got up and made their way down the cabin.

Two hours later, Vera emerged from her nap, her faculties fully refreshed. By this time, both Gussie and Fiona had succumbed to the extreme comfort of their fully reclining seats. The stirrings of Vera's considerable bulk behind the screen did not rouse them. Nor did the vibration on the cabin floor as she stomped purposefully past them. It was only when her indignant 'Well, this looks very cosy' resonated the along the length of the cabin, that they were awoken, with something of a start. When they peered round the cause of Vera's indignation revealed itself. Hen, now with her hair down, had her arm around Norrie Dick's shoulder and Winky was descending from Les's knee. Vera was not pleased.

'What's all this, then, Les Fyfe? I can't take my eyes off you for five minutes, can I? And as for you, Norrie Dick, Senga will have something to say about this. I'll trouble you young ladies to go back to your seats. I have something to say to my husband'. Vera proceeded to say what she had to say, volubly and at length.

Meanwhile, at the other end of the cabin, a debrief was going on.

'Well?' was the only question Fiona had to pose.

'What a creep! But he likes us. Thinks we're not nearly as stuck up as you'.

'Tell me something I don't know'.

'You did say that Les couldn't resist a boast. Well, he was going on about how he had made the Todd Institute and that they were struggling to get it off the ground before he came on the scene. He claims to have got it going single-handed'.

'Oh yes, he's very emphatic about that. He says that this guy Simpson is one of his former students and that no appointment is made at the Todd without Les's say-so'.

'I thought so. What's he plotting for this meeting, then?'

'Well', began Winky, they were talking about some troublemaker called Tom Carroll'.

Fiona blushed. 'Oh?'

'Do you know him?'

'Vaguely. The name's familiar. I believe I might have met him'.

'Les wants Simpson to get rid of him. Apparently he's a member of the Faculty at the Todd, and is being obstructive about the research that Les wants to encourage there. Les was complaining that this Tom Carroll was causing trouble at a conference in Estonia earlier this year'.

'Was that the same conference you were telling us about?'

'Must be, Hen, and I heard they had a big argument. But Tom, I mean Carroll has, I mean probably has, tenure. That would make him difficult to get rid of'.

'Tenure, yes he mentioned that, but Les says that won't save him. He thinks Joe Simpson is being much too weak and should get Carroll out of the way'.

'At least for the next six months'. Added Hen, with emphasis.

'Yes, he said he doesn't want anybody rocking the boat for the next six months', said Winky, taking back the topic.

'Who's Prunella Todd?' Hen didn't want to let Winky make all the running.

'Oh, I don't know, a member of the family that funded the whole thing, I suppose'. Fiona's thoughts were still taken up with the matter of six months.

'We'd worked that out! According to Les she is all powerful; both the Dean and the President are terrified of her and will do whatever she says. Les thinks she controls the purse strings'.

'She'll be at the Board Meeting and Les is going to have a word with her. He says she is very enthusiastic about Simpson's research and has asked if he needs more funding'.

'You look worried, Fiona'.

She was, but not for the reasons they thought.

'Interfering with tenure is really serious. It goes to the heart of academic freedom. It's almost impossible to remove someone who has tenure unless they have committed a crime, which Tom would never do'.

'How can you say that, you say you hardly know him. Anyway, Les seems to think there are ways round the tenure problem. Does Bear have a campus in Africa?'

'Yes, South Africa'.

'Well, Les was talking about the need for more research into treatments for tropical diseases. They all laughed'.

Fiona was now quite distracted. She would have to warn Tom, but what could he do? If Les turns the Todds against him, he'll have a very hard time. Of course, he could pick up a tenured position at a decent university tomorrow, but that would be very inconvenient. She needed him at Bear to keep an eye on Simpson and the office of the *Transactions in Moleculetics*.

'I think that was about all we could get out of him'.

'He's definitely got it in for Carroll and the others all said they'll support him completely.'

By now, Fiona was not listening at all. She would call Tom as soon as she got to Bear. It would not be a good idea for them to meet at the Guest House, as they had planned.

Les should not see them together.

'Oh, there was another thing. It seems that Les is going to be very rich. He was boasting about some stock options he has in a company called Pennine. Fiona?'

'What?'

'Les says he is going to be very rich. Some stock options he's got in Pennine'.

'Pennine? Probably nothing. Les has always got some get-rich-quick scheme on the go. Is that all?' Fiona needed time to think.

'I thought we did rather well! We are new at the espionage game, after all'.

'Oh, I'm sorry. Yes, that was very useful. Very useful indeed'.

'And Les has invited us for drinks at this fancy Guest House you're all staying at. He's promised us Krug'.

The remaining four hours of the flight passed without incident, both parties keeping to their own end of the cabin. Vera did not risk another nap. She was on guard duty. Guarding Les, primarily against himself, but also against the two floozies at the other end of the cabin. Her husband was getting up to his old tricks, and they were just the kind of students who had got him into trouble in the past. He would have to be watched closely. Les accepted his wife's vigilance quite passively, thanks to a combination of champagne, claret and brandy. In this, as in all else, his disciples followed, and the regular, low rumble that emerged from that end of the cabin reassured Fiona that she need not expect any conversation from them; reassured Paul and Hank that they would have a respite from the endless demands, and reassured Vera that she need expect no further breaches of security.

Winky, Hen and Gussie also slept, leaving Fiona to her

own thoughts, which kept her very much awake. So they were going to have Tom moved to Africa, were they? Of course, he wouldn't go. He would leave and take a position elsewhere. Then he would be free to speak out, and they wouldn't want that, would they? Nor would she; that could be very embarrassing for *Transactions*. What was so important about the next six months? Must be something to do with that work they are publishing in the journal. All the more important for Tom to remain at Bear, in that case.

An exquisite afternoon tea was served an hour before they landed. Paul, much to the amusement of Hen, insisted on calling it 'High Tea'. A solecism, but at 35,000 feet, an excusable one. Les insisted on champagne with his. As they approached the airport the passengers could see, to their general satisfaction, that it was a clear day in San Diego. The sudden greenery was a refreshing contrast to the endless, brown desert they had been traversing. This was much commented upon at both ends of the cabin.

The bureaucracy of landing: customs, passport control and baggage collection was smoothed by the Bear concierge service, who were on hand to greet the plane. They were soon at the appointed kerbside location marked 'Limousines', to attend the arrival of the two Bear limousines that were to take them to the campus.

'Where are the cars? They should be here by now'.

'The traffic's very heavy today, madam, but the cars will be hear any minute'.

'I hope so. Standing in the sun brings me out in blotches. I've got very delicate skin'.

Henrietta, Cordelia and Fiona studiously avoided eye contact at this point and managed to stifle their giggles. Limousines came and went: a constant line of Cadillacs, BMWs, Lincolns and Mercedes.

'Any minute, you said. We've been here at least five, and this sun's doing me no good at all'.

'Here they come now, madam' said the polite young man from the Bear concierge service.

Two enormous cars hove into view, the sun glinting on their highly polished radiator grilles.

'Rolls Royces?' squealed Hen.

'Bear has only Rolls Royces, madam'.

The Fyfes shook their heads and threw Henrietta a pitying look as they stepped into the first car. The party divided itself naturally into two. Fiona, although she was going to the Guest House rather than to Fairbanks Hall, went with the students rather than with the Board. The driver assured her that dropping her off at the Guest House would be 'no problem'. He had only two questions, and they were both for Fiona.

'Ma'am, what refreshment would you like to have waiting for you when you arrive at the Guest House. If you let me know, I'll call ahead on the car phone'.

'I'd order champagne, naturally. I bet Les Fyfe does', suggested Hen.

'How about a Manhattan? I love American cocktails'.

'No, Winky. I'm actually quite thirsty. Driver, a cup of tea will do nicely'.

'Yes, ma'am, what kind of tea?'

'Earl Grey, no milk, no sugar'.

'Of course ma'am. One more question'.

'Yes?'

'Would you like the services of our masseuse?'

'Masseuse? Are you serious?'

'I surely am. It's a great way to restore the circulation after a long flight'.

'Is there anything you people don't take care of?'

'No Ma'am. Welcome to the wonderful world of Bear'.

The two Rolls Royces purred north along the Interstate. In the first, silence reigned as its passengers slept. Even Vera slept, sure in the knowledge that Les could get up to no mischief here. In the second, the driver, Arthur, had ready answers for the many questions that his four eager passengers fired in his direction.

'Yes, Dr Hamilton, any time you want to go off campus, a Bear car is at your disposal'.

'Are they all Rolls Royces?'

'The four limos are, but we also have several SUVs – Sports Utility Vehicles – for rougher journeys'.

'Any time you want to go somewhere, just you give Arthur a call'.

'Thank you. That's very kind'.

'What about us plebs'.

'Excuse me, miss?'

'What about transport for students. I bet we don't get a Roller everywhere'.

'I'm afraid not, but as scholars you'll be able to use the car pool'.

Hen was mollified.

Winky wanted to know about shops.

'La Jolla has some very fine stores and restaurants also. If Dr Hamilton is interested, I'd be pleased to drive her into La Jolla'.

'And I'd need some shopping companions!'

'I thought you would, ma'am'.

The car phone rang. Arthur answered. He immediately sat up very straight.

'Yes, Miss Parsons, would you like to speak with her? Dr Hamilton, Miss Parsons for you'.

Fiona took the phone. The others, particularly Arthur, were very impressed.

Nods were exchanged and ears burned.

'Hello Monica'.

'Yes, I got it'.

'Martita said she had a fax from you'.

'It's perfectly charming, if you like pugs'.

'I'm sure she'll love it'.

'No, it's in my handbag'.

'Not at all. I'd love to meet you there, and it seems I've got a refreshing cup of Earl Grey waiting for me when I arrive'.

'No, nothing stronger. The champagne on the plane was very good'.

'I'll see you very shortly, then'. Fiona handed the phone back to Arthur.

'That was Monica Parsons'.

'Pugs?'

Her mother breeds them and collects pug *objets d'art*. I picked up a present for Monica in London.

'Well, Ma'am, if you're a friend of Miss Parsons, you're going to have a great time at Bear. Yes, she's a great lady, and you're going to have a great time. Yes, siree, a great time. She'll be giving you some of that ol' retail therapy. Miss Parsons is an expert in retail therapy. Yes, Rodeo Drive, here we come'. Arthur laughed and shook his head.

The Bear campus, bathed in the warm golden glow of a Californian evening sun, was every bit as exotic as the passengers of Rolls Royce 2 had come to expect. A landscape of immaculate green lawns, palm-lined drives, lakes and playing fountains was populated with a remarkable set of buildings. Here was a Greek temple, there a gothic tower, yonder a French chateau. But beyond and above them all, it shone, golden in the setting sun.

'The Todd' – four voices murmured in awe.

'It sure is, folks. That's the world famous Todd Pyramid. Bigger than Cheops. One of the great sights of the California coast'.

As the car swung round a corner the forest of lesser pyramids that clustered round the Todd came into view. The sunset sparkled on their glassy peaks.

'How many are there?

'Last count, six, but another two are on the way up'.

'Amazing, we should take a walk round campus tomorrow'.

'Walk? No need for that. You can use one of the Bear buggies'.

'Remember, Hen, people here walk for aerobic fitness, shopping and other recreational purposes, but never actually to get anywhere. For that you use wheels!'

'We got plenty wheels in this place. We sure do have plenty wheels. Yes, siree! And here we are, folks, Fairbanks Hall'. The car came to a stop in front of an enormous timber building with a thatched roof.

'Hall? It doesn't look like any hall I've seen'.

'Never been to Polynesia, miss?'

'No'.

'Folks tell me they got buildings just like this in Polynesia. Call them Tiki Rooms. But this is just the main building. You'll be staying in one of the lodges in the garden. Here's Wimea, she'll take care of you. Hey, Wimea, how you doin'?'

'Just fine, Arthur, just fine. Welcome to Fairbanks Hall', said the smiling lady in the grass skirt.

Fiona bade them farewell and Arthur started to the Bear Guest House. The phone rang.

'Hi there Dr Carroll, what can I do for ya?'

'Yes, she's here. Wasn't that a great game Saturday'.

'He's sure at the top of his form right now. Here's Dr Hamilton'.

'Tom. Thank goodness you phoned. You mustn't come to the Guest House'. 'Turn round then!' 'Of course I want to see you. I just don't want Les Fyfe to see you, at least not with me'. 'No, nothing like that'.

'I have some information about what's going on. I'll phone you as soon as I check in at the Guest House'. 'No, I can't now'.

'Yes, I promise, as soon as I check in'.

'Bye'.

'Thank you, Arthur'. Fiona handed him the phone.

'Dr Hamilton, you should have told me you were a friend of Tom Carroll. I'd have given you extra special treatment. He's one of the best'.

'That's quite all right, Arthur. You couldn't have been more helpful. You know Tom well?'

'Ever since he got to Bear. He was one of our best surfers. Coached my son Arthur Jr. Thanks to Tom, he made it into his college team'.

'At Bear?'

'No ma'am! He's at the University of California. I'd never let a child of mine become a student at …' But he broke off suddenly. 'We'll shortly be arriving at El Encanto, our Guest House. You can see it over there at the other side of the lagoon'.

Fiona turned and saw. After a drive through campus that had taken her on a tour of world architecture, from 'The Glory That Was Greece' via the 'South Seas' to the 'Tombs of The Pharaohs', she had been expecting something more exotic. El Encanto actually looked like it belonged in this landscape.

Arthur's 'We call it 'The Hacienda' confirmed this impression. 'It's the oldest building on campus. Parts date back 200 years, when the Spanish missionaries built it as a roadhouse for travellers along the coast. Of course, it's been extended quite a bit since then. Look, there's Simon, the head butler, waiting to greet you. And I do believe that's Miss Monica Parsons standing beside him'. The car came to a halt. Simon opened the door.

'Fiona, honey, welcome to Bear. I'm so delighted to meet you. How was your journey? You look pretty good, considering'.

'Considering what?'

'Considering I just saw your travelling companions. Compared with them, you look fresh as a daisy. It take it the bar on the plane will have to be restocked'.

'You could say that'.

'Honey, these guys are like a plague of locusts. They leave nothing. It's the same every year. The President wants me to keep an eye on them this year'.

'Good luck!'

'I'm off duty now, though. They've all crashed out in their cabanas, even the wife. Let's have that cup of tea you ordered'. Monica led the way to the Great Room.

'This is where the travellers of old used to bed down. We do rather better at Bear these days. It's so nice to meet you at last. Shall I pour?'

'Thank you'. Fiona opened her handbag and rummaged inside. 'I have something for you'.

'The dog?' Monica opened the small box. 'Oh, it's adorable. Ugly, but adorable. Moma will love it. Thank you so much. Now, let me tell you about your programme for the next few days'.

Simon reappeared. 'Can I take your coat from you, Dr Hamilton. I don't think you'll be needing it here'.

'Good idea Simon. Why don't you show Dr Hamilton to her cabana? I have to run a little errand for the President, but I'll see you tomorrow'. With that, Monica Parsons climbed into a golf buggy and headed off'.

On the way to Cabana Santa Barbara, Simon explained to Fiona that there were 15 such dwellings in the Guest House complex, each named after one of the original Spanish missions for which California was famous. He also pointed out to her the pool, tennis courts, spa and other facilities that were for the sole use of Guest House residents. With only one bedroom, 'Cabana Santa Barbara' was one of the smaller units in the complex. It did not command one of the famous ocean views, but overlooked the lagoon and had a small, private garden. Bridget, the maid, unpacked her case, hung her clothes, arranged her toiletries and eventually, to Fiona's relief, took away some things to press. Desperate to speak to Tom, she did not savour the sunset over the lagoon as she might have in more relaxed circumstances. She picked up the phone as soon as Bridget left.

'Africa! No way!'

'I thought you'd say that'.

'So that's what they're up to. Why six months? I don't understand that. They must be planning something soon. We need to get to the bottom of this. How about dinner?'

'No dinner, I ate plenty on the plane. Where can we meet?'

'Can you drive a golf buggy?'

'Never tried'.

'The butler can show you. This is what we'll do …'.

Fiona drove herself across the darkened campus towards the great, shining pyramid of the Todd. Tom was waiting for her by the reflecting pool in front.

'This is amazing. I take it that's the Windsurfing lagoon', said Fiona, pointing to the vast open space that occupied the lower part of the pyramid. All those sails are moving very quickly'.

'Evening training. But we'll go in the back way. Strictly speaking, I should check you in at the front desk, but in the circumstances, it's better that nobody knows you're here. We'll go straight up to my office. Let's park this buggy round back'.

They drove round to the staff entrance and took the elevator to the 8th floor. Like all first-time visitors to the Todd, Fiona was awed by the building and its facilities.

'I've never seen labs like these. The equipment. It's a dream'.

'Now you see why I put up with Simpson'.

They sat down in Tom's office. Fiona told him the full story of what had passed on the plane.

'But it doesn't make sense. The real threat to them now is Wiseman. He thinks that they are wrong and he's determined to prove it. He tells me he's actually been reading their papers in *Transactions* since he got back to Oxford and as he doesn't have his own research group any more, he's cajoling Sol Jacobson into putting one of his students on the case. Sol called me yesterday to complain. He says his group have got better things to do and that these chemical reactions they are reporting are not all that novel. But he's agreed to put a student onto this over the summer'.

'Fyfe won't know about this'.

'No, but he must be suspicious. Everyone knows how relentless Sir Henry can be when his curiosity has been piqued'.

'Where does he think you fit in – and why are the next

six months so critical? If the theory they propose is wrong, and Sol Jacobson's group is looking into it, an paper will be published that disproves it. Voila! That's what's known as the scientific process. It's happened before'.

'Yes, but there's a lot riding on this work here. The Todd family, who fund the Institute, have gotten very interested in it and are pumping in more funds. It seems Mrs Todd has a weight problem and thinks that this work could eventually lead to a miracle cure'.

'More money than sense, but that's nothing new. I can see why they don't want you here rocking the boat'.

'Maybe, but as I said, Sir Henry's planning to rock the boat big time. That's the real threat for them'.

'And Fyfe probably saw you speaking to Sir Henry in Tallinn'.

'Eliminate the spy in the camp? But I've got an understanding with ... You know, you're probably right. But I'm not going anywhere. I've got tenure'.

They talked for two hours, but despite the stimulating effect of Tom's special laboratory coffee, Fiona's jet lag began to kick in. Tom drove her back to El Encanto, where they bade each other good night and arranged to meet again on the evening of the Board Meeting. Thanks to the programme Monica had arranged, Fiona was fully booked until then. She made her way to 'Cabana Santa Barbara', where she fell into a deep sleep.

The pre-meeting programme commenced after breakfast the following morning. Fiona felt she had chosen the right options. Her all-too-intimate acquaintance with the Fyfes had taught her, among other things, that Les loved golf, but hated shopping or any sort of vigorous exercise. Vera loved shopping, but suffered from seasickness and also hated any sort of vigorous exercise. They both loved eating.

These predilections informed Fiona's choice of programme. Today she had arranged to go sailing, tomorrow she would take windsurfing lessons. Only on the third day would she indulge in a high-risk activity, shopping, but that would be under the supervision of Monica Parsons, who assured her that Vera Fyfe was scheduled for an all-day beauty session at the Bear health spa that day, in preparation for the Board Dinner that evening. 'Only one day? Will that be enough?' was Fiona's waspish response to this news.

The artificially-created wind was blowing the gentlest of zephyrs over the Todd Windsurfing Lagoon. It was a beginners' class, and Fiona was making good progress. Even Winky, Gussie and Hen, who had earlier been practising their advanced jibing techniques in a computer-generated Force 7, were impressed and shouted encouragement from the gallery. They all met up for lunch, and shared their experiences, so far, of the Wonderful World of Bear. They agreed that the real world would come as a great shock when they eventually returned to it.

'How are things with my friend Les?' asked Hen.

'I only see him at the dinners, and there are always lots of other people around, so I'm pleased to say that things with Les are fine'.

'What about the great plot? The one to get rid of your friend Tom'.

'My friend?' How had they found out? 'Oh, I mentioned it to him when I bumped into him at the Todd. We can't work it out. Why six months? He's confident that their theories will unravel anyway. I told you about Sir Henry Wiseman. Tom says he is on the case. It doesn't really matter what happens here. There must be something else'.

'Is he going to go to Africa, then?' asked Gussie. 'I'd jump at the chance. My grandmother told me such wonderful tales about her time in Kenya'.

'Don't be silly, Gussie. Of course he's not going to Africa'.

'Why not? Seems to me it's the last thing they'd expect him to agree to. Always do the unexpected, you might find something interesting. That's what granny used to say'.

'That's quite enough of granny!'

'Actually', said Fiona, 'perhaps Gussie has a point. They're probably hoping that he'll quit rather than go to Africa. Then he'd be out of their way for good. They'd never expect him to agree. That would throw them'.

'Clever old granny!'

'The other thing about Africa is that once you are there, it's remarkably easy to lose contact. Could be anywhere. Granny told me she was once up-country for two months and not a soul knew what she was up to.'.

'Yes, yes. Tom did say the real action will be taking place away from Bear. Once he's in Africa he could roam, even as far as Europe. They won't know. They just want him out of here. I think we have the makings of a plan. Hen, about your great friend Les …'

'Yes. What about him?'

'You said he invited you and Winky for a drink with him'.

'Yes, he did, but I'm not sure that his lady wife was so keen'.

'Tomorrow we have a free evening, to relax between the Board Meeting and the post-conference tour. Monica Parsons is taking me shopping'.

'Yes, but how does this concern Les?'

'Well, the one form of exercise Vera loves is shopping. Monica has been avoiding her like the plague, but I'm sure I can persuade her to allow Vera to join our expedition'.

'That's a big sacrifice'.

'I know, but it's in a good cause. It leaves the field free for you two with Les'.

'That's a pretty big sacrifice too'.

'England expects!' said Gussie in an outburst of patriotic fervour.

Monica Parsons was less than delighted at the prospect of an evening with Vera Fyfe.

'If you insist, honey, but I have always made it a rule to avoid the Fyfes whenever possible. They're such freeloaders, if you'll pardon the expression'.

'But it's in a very good cause'.

Fortunately, Monica was very sympathetic to Tom Carroll. As Bear's first PhD graduate, he could be deployed in media interviews when questions were raised about Bear's academic credentials.

'All right, if it's so important, I'll entertain Vera that evening'.

Monica also betrayed some misgivings about the way the Todd was being run.

'All that money pouring in, and what do they have to show for it? I keep telling the President, but he won't hear a word against it. Whatever Prunella Todd wants, Prunella Todd gets'.

The next day, at the Board Meeting, Fiona met Prunella Todd for the first time.

'Do I detect a Scottish accent? A fellow-countrywoman, how charming! You must come sit by me, I never understand half of what they're talking about'.

'Dr Hamilton, you can't sit there. You're here as an observer, not a Board Member. There's a seat over there against the wall for you',

Prunella Todd bristled.

'Professor Fyfe, Dr Hamilton is here as my guest and she'll sit wherever I want'.

'Certainly, Mrs Todd. No problem'.

The President of Bear opened the meeting.

'First on the agenda is a financial status report by the Treasurer'. The financial status of the Todd Institute was deemed to be very healthy indeed. The endowment income, combined with the revenues from the tours of the windsurfing facility, and regular, tax-deductible infusions of cash from the Todd family, meant a healthy surplus. Prunella had little interest in finance and chatted conspiratorially to Fiona throughout, much to Les's fury.

'Next, we move to the Director's report, from Dr Simpson'. Prunella now paid attention, whispering to Fiona. 'The research they are doing is SO exciting'. Prunella began to take notes. Simpson described the 'ever-more-promising findings' on the properties of the new class of chemical compounds being developed at the Todd. Fiona felt a sharp dig in her ribs as Prunella nudged her with her elbow. He also went on to show how the papers they had published in *Transactions* had been very widely cited. All the graphs pointed upwards. Fiona, tempted as she was to comment, did not forget that she was there only as an observer. Les's regular scowls in her direction reminded her of this fact. She said nothing. Prunella, on the other hand, had several questions.

'How are the rats doing?'

'The initial weigh loss has been sustained'.

'Have you tried it out on any other species'.

'Not yet, ma'am'.

'Well, I think you should. This work is meant to be for the benefit of mankind'.

Les Fyfe interjected. 'Mrs Todd, we are confident that

this material can be made available for human therapy within a short space of time. All the tests so far are very positive'.

'On rats. I know it can take years to get new drugs onto the market. Isn't it about time you moved beyond the rat stage'. Much as she was tempted to do so, Prunella did not mention that her own experiments, which had moved onto the Labrador stage, had been very successful. 'Shouldn't you be moving into clinical trials?' Prunella had been reading a *Time* magazine special on pharmaceuticals.

'Not necessarily', said Joe Simpson, but a sharp look from Les prevented him developing this point.

'That's enough, Prune. They know what they're doing'. Franklin Todd did not want the subject of testing on beagles to come up again. He had made his views on this very clear to the President in private.

Simpson moved onto the next slide. 'Health Africa' is a new initiative. We want to set up a special Todd unit at the Bear Africa campus to research the therapeutic properties of the unequalled range of naturally occurring chemicals found there. Science has hardly scratched the surface ...'.

Prunella Todd was entranced. Any talk of Africa brought back memories of her darling Frankie and the bravery of those wonderful trackers who tried to drive the enraged buffalo from him. The word 'natural' also struck a special chord. Prunella Todd was a firm believer in natural remedies and organic foods.

'We plan to send one of our brightest scientists to Bear Africa to set up a special unit for the investigation of natural products from plants. It will not cost a great deal of money, as you can see, and we would like the approval of the Board to do it'.

The 'definitely approved' shouted by Prunella was

music to Les Fyfe's ears.

The 'This is very worthwhile' she whispered to Fiona was less warmly received.

Prunella and Fiona were both silent for the rest of the meeting. Prunella absorbed in plans for yet another memorial to her lost son. Fiona thinking how she could persuade Tom to take the position. The more she thought of Gussie's idea, the more she liked it.

'As they enjoyed the cocktail, traditional at the end of the Board Meeting, everyone seemed very happy with the way things had gone. Les was loud; his plan had worked. The President beamed; he scented more Todd money. Prunella glowed; she loved having another opportunity to commemorate her Frankie.

'Fiona, it's been so nice talking with you. How long will you be in California?'

'Another five days? Then you must come to see us in Montecito. How about lunch?'

The obstacle of the post-conference tour was overcome. The President, always eager to fulfil Prunella's wishes, would arrange for Fiona to leave the winery a day early and fly down to Santa Barbara. The date was set. Prunella Todd, much to the disgust of Les Fyfe, gave Fiona her private phone number, and with an air kiss to both cheeks, departed with husband and President in tow for the Todd Windsurfing Museum awards dinner.

Fiona, not being able to avoid Les on her way out, made the most of it.

'Les, how about dinner? We seem to have a free evening'. Fiona knew he had other plans.

'Oh no, no. I can't. Er, Vera wants me to rest after all these meetings. I'm going to have a quiet night in the cabana. She's going shopping with Monica Parsons, you know'.

'Too bad', said Fiona as she left the room. She wondered exactly what sort of quiet evening Winky and Hen had planned and steeled herself for her own evening in the company of Vera Fyfe.

She heard Winky and Hen before she saw them. Their shrieks of laughter, as they bade Les a very voluble good night, echoed through the grounds of El Encanto. She had come back early to meet them. Monica had, under protest, agreed to take Vera for a drink in La Jolla's most sophisticated piano bar, and to keep her there until 10.30, by which time Les would be safely back in his cabana. It was now 10.05 and the approaching laughter was suddenly silenced. Winky and Hen were approaching Cabana Santa Barbara, and knew they had to do so under cover. Fiona barely heard the knock on her door.

'Quick, inside'. The door opened and closed swiftly.

'What a creep!' were Hen's first words.

'I think I've got bruises', were Winky's.

'All in a good cause. What did you find out?'

'Drink first, information after'. We didn't touch a drop. Had to stay on the ball, so to speak.

Fiona produced the promised bottle of Krug from the refrigerator and poured them each a glass.

'Yum-Yummy. Les is very pleased with himself. He said he got the Board to agree to send Tom Carroll to Africa'.

'I know that. I was there, remember. But what's his motivation'.

Hen sipped thoughtfully, as Winky chimed in.

'He mentioned again that the next six months are critical'.

'Did you ask why'.

'I told you before, there's this company called 'Pennine ...'

'No you didn't'.

'Yes I did. On the plane. Les Fyfe is going to be very rich and its got something to do with this Pennine company'.

'In England', added Hen 'and the next six months are critical'.

'Yes, I've got that point, thank you, but I still don't understand it'.

'If you give me some more champagne, I'll explain'.

'Pass over your glass, Hen. But what's the connection with the Todd'.

'Please do bear in mind that we couldn't get Les very drunk. He just wouldn't knock it back the way he did on the plane'.

'I think he had hopes … for later on', murmured Winky, with a grimace.

'He just let slip that 'we'll get some of the Todd guys over to Pennine as soon as Carroll sets off for Africa''.

'Then he started singing 'Who wants to be a millionaire?'

'What about the research going on at the Todd?'

'The stuff?'

'Yes'.

'Wink tried to get it out of him'.

Cordelia summarised the exchange between them

Me: 'You told us it'll be the answer to every woman's prayers'.

Les: 'Did I?'

Me: 'You told us it will make us all eternally young'.

Les: 'Did I? I don't remember that', he said. Then he changed the subject very quickly and started telling us how Vera doesn't understand him. And you know where that leads'.

'Just as well there were two of us'.

'After that it got very boring, so we reminded him we had to be up early for training tomorrow'.

'The drive back was not pleasant'.

'Poor Wink, stuck in the back seat with Les. I'm so glad I said I'd drive!'

'Fiona, are you listening?'

'Pennine? Pennine? I've never heard of them. It must be some kind of drugs company. I'll need to do some research when I get home'.

'Fiona? Please don't ask us to try to pry any more secrets out of Les. We've both decided espionage isn't for us, haven't we Hen'.

'Yes, a most disagreeable occupation. I'm black and blue all over'.

'Of course, you've done a great job. Les is a wily old fox. This Pennine connection sounds significant. Now I've got to persuade Tom to go to Africa'.

'Gussie's plan? He will be chuffed. More champagne? When do you talk to Tom?'

'He's coming here tomorrow, for breakfast'.

Winky and Hen exchanged significant looks. 'Should we be going?'

'I said 'for breakfast'. Really!'

'OK, no need to get shirty with us. Methinks the lady doth protest too much'.

'I don't know what you can possibly mean'.

More significant looks. Then a rather firm, masculine knock on the door.

It gave all three a start.

'Who's there?'

'Good evening, Dr Hamilton. Simon here with your dry cleaning'.

'Oh, Simon, just one moment'. Fiona sprang to the door,

which she opened a crack.

'Thank you Simon'.

'Will that be all, ma'am'

'Yes, thank you'.

'Winky and Hen showed no signs of going. They had told Fiona all about their evening and the balance of trade in gossip required that she should tell them something of her day. They wanted to hear more about the Board meeting, but saw they would have to fish.

Hen cast the first fly. 'They discussed Tom Carroll being sent to darkest Africa at the Board Meeting, then?'

Fiona could not resist the bait. 'Yes, they did, and Simpson sold the Todds on it very cleverly. You know their son died there'.

'We know, That's why we're all here. A toast to Frankie!'

'When he was trampled by the buffalo – it seems it was all his own fault, got too close – the trackers with him risked life and limb to draw the buffalo off. One got badly injured. Consequently, Prunella Todd goes misty-eyed whenever Africa is mentioned, Noble Savage and all that sort of thing. She's been out there several times and is always looking for ways to help them. Simpson knew that he would get instant support for his African 'initiative' to set up a branch of the Todd there to investigate natural products'.

'What are they?'

'Well, they're the chemicals that occur in nature, that you can extract from plants, etc. Lots of them have medicinal properties. So Simpson bangs on about the benefit to mankind ... South Africa has the most diverse plant life on earth ... needs more study ... want to send one of our best scientists out there ... he was very clever about it.'

'Sounds like he pressed all the right buttons'.

'He certainly did, and the outline plan for Todd Africa was unanimously approved. A sub-committee has been set up to organize the funding. Joe Simpson was told to speak to Tom tomorrow'.

'Just as well you're having breakfast with him, then. You can tell him all about Plan Gussie'.

Next morning Tom Carroll arrived at the Cabana Santa Barbara at 8.00 a.m. sharp. He had already spent a productive hour in the lab. Simon had just finished setting up breakfast on the private deck that overlooked the lagoon. Fiona emerged from the living room just as Tom took his seat.

Simon poured the coffee. 'The hot breakfast's in the trolley. Anything more I can get you, Dr Hamilton?

'Thank you, Simon. That will be all'.

'Call me if you need anything else'. Simon left.

'You sorry to be leaving Bear?'

'It's very seductive, but it's time to get back to the real world'.

'The real world of post-meeting winery tours?'

'One has to re-acclimatise gradually. It's very rarefied here. You can't just expect your guests to be sent back into the real world in one step. It's very tough out there'.

'And what is your real world, Fiona? We're talking about publishing, aren't we? From what you've told me, I don't think that will take too much acclimatisation from here'.

'Very funny', Fiona struck back, 'anyway, from what I heard at the Board meeting yesterday, I'm not the only one who'll be leaving Bear'.

'Oh yeah?' Tom put down his spoon. Fiona took the bull by the horns.

'Yes, Joe Simpson described the plan for the new branch

of the Todd at Bear Africa and said how they wanted to send one of their best scientists to get it off the ground'.

'Me?'

'Of course'.

'I won't go. They can't make me. I've got tenure'.

'The problem is that Prunella Todd loved the idea. I think she would take a very dim view if you declined this opportunity. And you know that if she takes a dim view, President Winthrop will also take a dim view. A promising career could be blighted'.

'I'll quit. I could walk into a tenured position at Yale tomorrow'.

'And how is the surf on Long Island Sound? Face it, Tom, you don't want to leave this place'.

'There's only so much abuse I'll take. Fiona, I didn't think you were so dumb. You know they only want me out of the way to leave them a free hand with this crappy line of research. I'm going to talk to Winthrop himself about it'.

'You know what his answer will be. Prunella Todd is 100% behind this plan'. Fiona decided to play her joker. 'As a matter of fact, I think that going to Africa is the best way you can scupper Fyfe and Simpson'. That silenced him. He looked at Fiona, open mouthed. When he could bring himself to speak, it was only to say that he was stunned, as well as outraged that she should suggest such a thing. Fiona told him to listen and proceeded to describe Gussie's plan, without attributing it, in so many words, to Gussie. She reinforced Gussie's strategic arguments with the facts that had been gleaned by Hen and Winky on the previous evening without specifically mentioning either of them by name. Fiona proved a powerful advocate and by the end of her exposition Tom's protests had faded. He had many questions.

'Pennine? No, I've never heard of them'.

'Nor me. They're based in England and there is some relationship with Fyfe and Bear'.

'How long to fly from South Africa to England?'

'About 12 hours, I believe, but no jet lag, as they're on the same time zone'.

'Do you really think I could just disappear while I'm in Africa? Make trips to England without them knowing?'

'Look, as far as they are concerned, you'll be on long expeditions up country, gathering exotic flora'.

'Yes, you're right, if we go far enough up-country, the phone and fax probably won't be too reliable'.

'And, you'll nced at least one trip to Kew'.

'Kew?'

'The Royal Botanic Gardens in London. They know everything about exotic plants there. You'll need to consult them'.

'Of course. I can see I'll need a big travel budget, but I'll be in a strong negotiating position. I could probably ask for what I want'.

'You should probably take an assistant as well. Someone you can trust'.

'A trusted assistant? Good idea. I could take Tai-Wah Chung, he's good; or Susan, she's the blonde you met the other day'.

'Tai-Wah Chung, definitely'.

'Tai-Wah? Why so definite? I didn't think you knew him'.

'I don't. I just know that Africa can be a very, very difficult place for women. Especially blondes'.

I suppose you're right. I probably shouldn't take a woman there'.

By the time they had eaten their pancakes, Tom Carroll

had come round to the idea that a sojourn in Africa would not only benefit his career prospects at Bear, but would also be the best way of finding out exactly what Fyfe and Simpson were up to.

'We should have someone watching things here as well. What about Susan?'

'Susan? Oh no, much too obvious', said Fiona, with feeling.

'Yes, you're right. Joe would be suspicious of her. Too close to me'.

Fiona had a brainwave.

'Winky and Hen? Those gals? Are you sure? What do they know about science'.

'Not much, but they're very good at extracting information from men of a certain age. If it hadn't been for them …', but Fiona decided she had said enough about their qualifications. Tom was already well aware of their principal attributes.

'Tom, you find some reason to introduce them to Joe. He'll have heard all about them from Les, so I bet he'll jump at the chance of a meeting'.

'True, nobody would suspect them. OK, if you say so. I'll make sure they meet Joe very soon'.

'And another thing. Don't seem too eager to accept. That'll make him suspicious'.

'I'll play hard to get. I'll want money for travel – plenty. And an assistant. As well as equipment'.

'Excellent! He'll love that'.

'So I'm off to Africa, then. That wasn't the outcome I planned when I agreed to meet for breakfast, Fiona'.

'Well, Gussie's granny always said to do the unexpected'.

'What's Gussie's granny got to do with it?'

'Oh, nothing. It's an old Scots saying'.

'Fiona, I don't think you're over your jetlag yet! I'd better hit the road and let you get ready for the Napa Valley. Remember, dinner on Thursday. I want to have at least one evening with you before you go back to the old country. Thanks for breakfast. ' Bye, Tom. Till Thursday. Good luck'.

Tom picked up his bicycle and sped off around the lagoon towards the Todd. Ten minutes later he called.

'I'm summoned to see Joe Simpson at eleven'.

'So soon? Too bad', Fiona said. 'I'll be on the plane by then. 'I'll phone you this evening to find out about it'.

She called Winky and Hen, but they were both out windsurfing. She left a message with their concierge to let them know that she would call back that evening. She now steeled herself for two solid days in the company of Les, Vera and friends. 'Thank God for Prunella Todd', she thought to herself. 'I escape one day early'.

Chapter 11: Another Luncheon Party

The Todd limousine was waiting to meet Fiona at Santa Barbara airport. As they wound through the tree-lined lanes of Montecito she had tantalising glimpses of the enormous mansions for which that community was famous.

'Big places round here', she observed to the chauffeur.

'Miss, they ain't nothin' compared to Elysium – that's Mr and Mrs Todd's place higher up in the hills'.

Indeed they were not. As the gates of Elysium opened, and the security guard saluted, Fiona saw what the chauffeur meant. The enormous Spanish-style edifice sat atop a ridge that commanded a panorama of meadow, forest and mountain. On a clear day it would also have taken in a substantial part of the Pacific Ocean, but today was one of those summer days, all too frequent, when the California coast was shrouded in low cloud. Elysium, true to its name, floated serenely above this, bathed in sunshine. As the car drew under the porte cochere, Fiona saw a large woman emerge from the house and take up position behind the dapper little butler who opened the car door. His welcoming

'Good morning, madam' was curtailed as the large woman stepped forward, saying, 'You can leave Dr Hamilton to me, Desmond'. She evidently spoke with some authority, as Desmond immediately stepped back with a slight bow.

'Welcome to Elysium, Dr Hamilton. Mrs Todd is looking forward to seeing you again. I'm Mitch, by the way, Mrs Todd's general dogsbody. You can follow me'.

'Thank you. Good morning'. She smiled at Desmond as she followed in Mitch's considerable wake'.

'I'm sure you'll want to freshen up before lunch. I'll take you to a guest room. Mrs Todd tells me you're a fellow-Scot'.

'Absolutely. I thought you sounded Scots, Mitch'.

'Och, aye. I've been here a quarter of a century and I'm told I sound like I've come straight from The Mearns. Mrs Todd likes me to lay it on a bit thick for the visitors, but I am the genuine article.

On the way to the guest room Fiona and Mitch exchanged credentials in the way that Scots who meet abroad invariably do.

'Really! I have cousins there!'

'Such a small world. My brother's wife works there'.

'An auld auntie of mine was born in that very street'.

'Oh, yes, a lot of changes'.

'I knew his father'.

'Here you are, Dr Hamilton'.

'Please, Mitch, do call me Fiona'.

'That's very nice of you Fiona, but not in front of Mrs Todd, if you don't mind. She likes us to maintain standards'.

'Was that her dog in the hall?'

'No, that's Missy, she's the apple of Mr Todd's eye. Spoilt rotten. Mrs Todd prefers Pekes, so watch where you sit down; they like to sleep behind the chair cushions. One got badly injured when Orson Wells came to visit. He got hurt too, mind you, needed stitches. Pekes have got sharp teeth, the little beggars. Missy's worth the whole pack of them'.

'Pack?'

'Mrs Todd has got four. She picked up the Peke habit from her mother. About the only thing she picked up from her mother, more's the pity'.

From these terse utterances Fiona was able to capture the essence of Mitch's opinion of Missy, the Pekes, the Pekes' mistress, and the Pekes' mistress's mother. Mitch, Fiona realized, could pack a lot into a few words.

Mitch left Fiona in the La Cumbre bedroom, which commanded a view of the peak of that name. She now almost regretted declining the invitation to stay overnight; but she was eager to see Tom before leaving for home.

After she had freshened up, Fiona rang the bell, as instructed by Mitch. Desmond appeared.

'Ready for luncheon, Madam? Please follow me'.

She followed Desmond at a stately pace down the stairs, through the hall, across a large room, which she later learned was the 'Saloon', onto an awning-covered terrace that overlooked the cloud shrouding the coast below.

To the right a table was set for lunch, while to the left Prunella Todd was seated on one of several armchairs. She appeared to be deeply involved in an energetic exchange with Mitch'.

'No, no, no, Mitch. I'm not taking the big stuff if you're not travelling with me'.

Prunella made very definite gestures with her aperitif glass.

'But Lady Maitland will be expecting it'.

216

Desmond coughed discretely.

'I don't care'.

'Frida is quite capable of looking after them. You know I have to be here when Geordie and his family arrive. They're flying 8000 miles to see me'.

'Of course, you must be here, Mitch. I'm not complaining about that, but I'm not comfortable taking the big stuff when you're not with me. Remember what happened in Delhi. I can borrow one of mummy's tiaras'.

Desmond coughed again. They both looked round and their frowns turned to smiles when they saw Fiona.

'Fiona, darling. Welcome! Do sit down. What would you like to drink? Desmond, a gin and tonic for Dr Hamilton'.

'Mitch tells me that you know some people in common. Isn't that typically Scottish. It so happens I'm flying there next week for a ball. One of my nephews has decided to get married. My mother is insisting on full tenue'.

This prompted Mitch to resume battle. 'You won't be the only one wanting to borrow one of her ladyship's tiaras. They might all be spoken for already, and you've got plenty of your own'.

'Mitch, I'm beginning to find this discussion very tedious'.

'It's up to you, but you know what Her Ladyship can be like'.

An image of what Her Ladyship can be like sprang into Prunella's mind. It was not a pretty picture.

'Oh, very well, but just the little one with the pearls'.

'And the matching parrure?'

'Yes, very well. But no emeralds. And remember the Maitland sash'.

Mitch rolled her eyes, but remained at her station, arms folded, ready for further dispute with her mistress, who had other ideas.

'That will be all, Mitch. I'll need a cocktail dress for this evening. The Reagans. You know'.

Mitch knew all her mistresses movements very well, and resented such a pointed reminder. She turned on her heel, and made her way reluctantly into the house, muttering to herself as she went. Although dusting the drawing room was not, strictly speaking, within her job description as 'personal maid' she decided that the collection of Meissen figures on the tables by the French windows leading onto the terrace were in need of her attention. Mitch might be advancing in years, her back might ache, her feet swell and her eyesight might not be what it was, but her hearing retained the acuteness of a rottweiler on guard duty. As she pretended to dust, Prunella and Fiona finished their drinks and moved to the luncheon table. Both ladies accepted Desmond's offer of white wine and tucked into their salads. They finished their pleasantries about Scotland with the first course. Mitch had been sorely tempted to correct some of her mistress's more outrageous errors of fact, but, with great effort, managed to maintain silence. The second course arrived and with it an offer of red wine from Desmond. The white had been so good that Fiona was tempted to continue to a third glass.

'Not with venison, darling'. The red wine was accepted and enjoyed by both ladies.

A second glass of red wine reinforced Fiona's Dutch courage.

'Fascinating work Joe Simpson and his group are doing'.

'Yes, isn't it wonderful. We're so proud to be associated with it. It'll make such a difference to so many people's lives. There is so much temptation out there, a treatment that suppresses our appetite is just what we need. My friends are all so excited about it. Such a pity it takes so long to bring it to market. All those ridiculous regulations. President Winthrop tells me it could take 10 years. *Ten years*! I'll be an old lady by then'. Prunella emphasized the point with her steak knife.

'Yes, but they do need to test these things thoroughly. They can have very bad side-effects'.

'The rats all seem perfectly fine. The ones on Compound X all look like supermodels'.

'Compound X?'

'Yes, that's what they all call it'.

'But rats aren't humans. They need to do tests on other species, and then they need properly controlled clinical trials. It all takes time'.

Prunella leaned closer to Fiona. 'Actually, darling, Compound X has been tested on other species'.

'No! They didn't mention that in the presentations'. Fiona had suspected they were not being given the full story.

'That's because they don't know about it'.

'What? How could they not know about it'.

'That's because I'm doing the tests myself'. Mitch almost dropped a Meissen monkey figurine at this revelation.

'Oh, Mrs Todd, you shouldn't! You could do yourself a lot of harm'.

'Not on myself, silly, on Missy'. Mitch's sigh of relief fortunately sounded like a gust of wind on the terrace.

'Missy, my husband's labrador. Much too fat. I've had Mitch give her Compound X the past two months and she's so much thinner. Never looked healthier. It works, I tell you. Professor Fyfe said there is a company in England very interested in bringing Compound X to the market. Called Pennine, I think. He said we should buy shares now, before they go up, but Franklin won't let me. Says it would be insider dealing. Spoilsport.'

At this point Mitch, who was finding the conversation ever more intriguing, collided with a table sending a lamp crashing to the floor.

'What's that? Mitch, is that you? She's always doing this'. Prunella leapt up. Mitch decided to make herself scarce. Fleeing toward the hall she grabbed a slumbering peke from behind a cushion and tossed it in the direction of the fallen lamp. As Prunella strode into the room she was greeted only by a rather bemused little dog.

'Oh Yum-Yum, it's you! Have you been a naughty girl? Any cuts, darling?'

Prunella re-emerged onto the terrace with a pekinese underarm. Desmond, who had heard the crash, was already there, full of concern.

'Desmond. Yum-Yum seems to have collided with one of the tables in the drawing room. Can you have the pieces cleared up and make sure that Ko Ko and Pooh Bah are OK. I don't want them to cut their little paws. Oh, and keep Missy away too'. Desmond went in search of a parlour maid.

'Fiona, have you met Yum-Yum? That was a big table for a little dog. I didn't know pekes were so strong'.

'Is she OK?'

Prunella checked each of her paws with great care. 'Oh, yes, just a bit shaken, aren't you darling. Where were we?'

'We were talking about Missy's weight loss'.

'Missy. Yes. She's definitely much thinner and much healthier'.

'How much of the stuff do you give her?'

'I leave that to Mitch. We did some calculations, based on the rats'.

Desmond reappeared. 'Dessert, madame?'

'Of course, Desmond. You'll have some crème brullee, won't you, darling'.

'Yes, please. It's one of my favourites'.

'Oh, you'd like some too, would you Fiona? Desmond, be sure to break the crust on Yum-Yum's crème brullee'.

'You seem very excited about the African project'.

'Yes, very excited. I think they can do wonderful things out there'.

'Joe seemed to have someone in mind to lead the team'.

'Yes, they told me afterwards. I can't remember his name, but they phoned to say he had accepted. It's going to be expensive, but we'll channel some of the profits from the windsurfing Institute in that direction. Anything to help these wonderful Africans. My Frankie would have been so proud'.

After dessert they moved back to the armchairs for coffee. By this time, the cloud had lifted and Fiona saw for the first time the spectacular panorama that took in a great sweep of coast, the city of Santa Barbara and the offshore islands. Conversation became more spasmodic as the wine and the afternoon heat began to take their toll. Prunella and Yum-Yum nodded off. Fiona excused herself, without disturbing them. She went in search of Mitch. If she could get her hands on some Compound X she could have it analysed, identified and properly tested. Mitch was nowhere to be found, but Desmond was. 'Gone to town to do some shopping. She said she'll be away all afternoon'. Following her collision with the table, Mitch had decided to make herself scarce; she also had an urgent matter to discuss with her stockbroker.

'I'd better think about leaving myself. I have to catch a plane to San Diego at five'.

'Shall I order the car, miss?'

'Yes, please, Desmond. I'll go and take my leave of Mrs Todd'.

Tom Carroll met Fiona at San Diego Airport. Instead of driving her straight to Bear, he suggested they take a detour to Del Mar, where they could have a bite to eat in a beachside restaurant that was a favourite of his.

'Welcome to Neptune's Locker. A party of two?. Follow me'.

The intimidatingly lean, tanned blonde led them to a table that was right on the beach, and took their drink order.

'This is one of my favourite spots in the world, especially at this time of day', said Tom. 'Just look at that sunset '.

Fiona looked at the reddening sky, at the burnished gold of the ocean, at the silhouetted couples strolling along the beach, and at Tom, who's gaze was fixed on some point far beyond the horizon. She was going to miss him. For once, she was not looking forward to going home, and she knew why.

Their drinks arrived, and they sat in silence until the last rays of the sun fell from the sky. Tom turned towards her.

'I'm going to miss you, Fiona. I've enjoyed the times we've had together'.

'Me too'. She was relieved that the enveloping darkness would hide her ferocious blushes. 'But we'll meet again, when you're in 'darkest Africa''.

'Yes, when I'm in 'darkest Africa'.' They both laughed. 'Too bad you couldn't get your hands on Compound X. We could have done some tests'.

'I know, but I'm sure that Mitch suspected I would come looking for her and took evasive action. Prunella Todd wouldn't believe me when I told her Mitch had gone shopping. 'Mitch hates shopping. Refuses to do it', she said, and called for Desmond – he's their butler – to confirm my story. Yes, Mitch was definitely avoiding me'.

'And the dog was thin, was it?'

'I had to take Prunella's word for it, but Missy did look thin for a lab'.

'Not as thin as that peke that Orson Welles sat on, I bet!'

'Very funny. I thought you liked dogs'.

'I do, but pekes? They don't count!'

Their conversation turned to Africa.

'So you've definitely agreed to ship out to Africa in September?'

'Yes, I can't go before then. I've got conferences to attend and students who are finishing up their research projects. Simpson wasn't too happy about it. Wants me out in July, but I said 'No Way'.

'And the money'.

'All the travel money I want. They'll wire funds to me. He even said I can fly out there first class'.

'Via London?'

'Is there any other way?'

'He wants weekly reports, but that won't be a problem. Tai-Wah can fax them for me'.

They arranged that he would stop over in London for a few days on his journey to South Africa, and that in the meantime, Fiona would find out what she could about Pennine.

'Well, I'd better be getting back to El Encanto. I'm meeting Winky and Hen there later'.

Simon was waiting at the main door of El Encanto, pointing out to a driver of a large truck that the delivery entrance was round the side. He told them to hurry along when he saw Tom's car pull into the drive.

'Dr Hamilton, welcome back to El Encanto. So nice to see you again. There are two messages for you. Miss Young would like you to call her as soon as you check in. And Ms Parsons called to confirm your breakfast with her tomorrow morning'.

'Oh, I'd forgotten about Monica. Sorry, Tom'.

'Never mind, we'll always have 'darkest Africa'. Remember, I'm taking you to the airport tomorrow. I insist'.

'Thank you, Tom, you're so sweet'. She gave him a peck on the cheek before following Simon to Cabana Santa Barbara. It was Tom's turn to blush and he did so profusely as he climbed back into his car.

'We've been so clever', boasted Winky as she breezed into the cabana.

'Yes, we managed two evenings with Joe, and got him drunk both times. We found out that *his* wife doesn't understand *him*!'

'I hope you can do better than that'.

'We might, with the right fuel'.

'What?'

'Krug, darling'.

'Fiona opened the refrigerator'. No Krug.

'Oh dear, it looks like they haven't restocked the fridge'.

'No Krug, no story. That's the way it is. Can't you call for some. I thought this place was awash in the stuff'.

Fiona called Simon. Things did not look good.

'I know they drink a lot, but not the entire contents of the cellar, surely!'

'Yes, their trunks did seem very heavy. They needed porters wherever they went'.

'All the Krug and the Chateau Margaux. They go for the best, don't they'.

'A new shipment? Could you check? An hour to chill? That would be great. It's an emergency. Thanks for your help, Simon'.

'OK ladies. It seems that Les and his pals have helped themselves to the cream of the El Encanto wine cellar. Simon says they were ordering vast quantities of Chateau Margaux and Krug every night, but left hardly any empties. He suspects that they stuffed their baggage with them. I wondered why they were all travelling with huge steamer trunks. The good news is that fresh supplies have just arrived and the Krug is chilling. Simon's on his way over with a bottle of Pol Roger to keep us going. Now, spill the beans!'

Two hours and three bottles of champagne later, Fiona was in possession of some very pertinent information.

On their first evening with Joe, he was still flushed with his triumph over Tom Carroll, who was to be 'exiled' – his words – to Africa. Tom was a 'troublemaker' who was intent on spoiling one of the greatest scientific discoveries of the century. 'The key to weight loss', was how Joe had put it. 'As soon as he's out of the way, they'll be able to step up production of Compound X. The necessary equipment was all ready to be installed'. He stressed that everything had to be done in the utmost secrecy, to prevent possible competitors knowing what they were up to. They were going to 'go commercial' with it and had a partner in England. This was because it would take much longer to get permission in the US to do what they want. At this point, Joe became melancholy and had lamented the fact that his wife did not understand him. His collapse into a tired and emotional state brought an end to their first evening's interrogation.

The second evening had proved much more difficult. Joe was unwilling to be plied with quite so much drink, and clearly had other entertainments, involving both Winky and Hen, in mind. Hen, cornered by him on a sofa while Winky was out of the room, decided that extreme measures were needed.

She had blurted out 'Les told us all about Pennine', which produced an instant mood swing. He had leapt back like a scalded cat, hissing repeatedly, 'He never would, he never would. You're making this up'. He had been loudly accusing Hen of being a liar when Winky walked back into the room. He insisted they tell him what they knew about Pennine, and having concluded that they knew very little indeed, warned them to stay away from the Todd laboratories and bade them a frosty good night.

'It was very scary', said Winky.

'Complete personality change', added Hen.

'We don't want to see that man again', shouted both.

Fiona calmed them down and promised that no more 'Miss Marple' would be required. After many hugs and kisses, with promises to get together when they were all back in England, Winky and Hen left Fiona to do her packing. Now alone, there were but two thoughts on her mind: Tom and Pennine.

Chapter 12: Fiona Goes North

On her return to England, Dr Fiona Hamilton was busy on a number of fronts. First, she submitted her Trip Report, covering events in California to Henri de Klompenmaker. Scientifically thorough in these matters, she gave an exhaustive account of the official part of her programme, but said nothing of the extracurricular meetings that took place in the plane, limo, or Jacuzzi pool. She had, as yet, little hard evidence to support her suspicions of improper conduct by the Editor *ad interim* of the *Transactions in Moleculetics*. Until she had this, discretion would be the better part of valour. A few days later, she called Klomp's secretary to check that her Trip Report had arrived. Yes, it had, but he had been very busy and had not yet read it, and no, he would not be able to meet her this week to discuss it.

'Most unKlomp-like', she thought as she put down the phone. 'Anything to do with *Transactions* usually has top priority. In less frantic times, Fiona might have dwelt on this unusual behaviour, and might even have set up an informal seminar with some fellow-Klompenologists to discuss it, but not today. She had bigger fish to fry.

She phoned Sir Henry Wiseman, to let him know that

Tom Carroll would be passing through London in October. Sir Henry had expressed his delight at the news, and mentioned that Sol Jacobson would also be in England at that time. He informed her that Sol's group were making good progress on research that would refute the claims made by Les Fyfe in Tallinn for the Todd work. Sol was coming to Oxford to confer on an paper that they would jointly author for *Acta Moleculetica* on the subject. Fiona and Tom were bidden to dine in Hall at St Leonard's College as Sir Henry's guests.

Fiona Hamilton then turned her attention to Pennine. She had been itching to look into this and her first port of call was the SI Market Research Department. They did not disappoint. A search of the Companies House database, where every British company is registered, revealed that there was indeed a company called Pennine and that one Leslie Fyfe was a Director. But Pennine Nutritional Supplements, based in Rochdale, was not in pharmaceuticals. It produced dietary supplements for chickens and other livestock. This both excited and frustrated Fiona. She was excited that she had found Pennine and had made the connection with Les; frustrated that she could not work out what an animal nutrition company was going to do with a chemical that caused weight loss. She knew little of animal husbandry and thought that the point was to fatten them up, rather than thin them down. She would have to speak to an expert. Having printed out as much information on Pennine as she could, she thanked the market research department, made her way back to her office, and faxed all 45 pages to Tom Carroll's private number in California. Over the following few days, she had regular phone consultations with him, in which they debated where the relationship between Les Fyfe, the Todd and Pennine might be leading. As far as Les was concerned, it was going to lead to great riches. This was the clear implication of his drunken boasts to Hen and Wink. But how? 'There just isn't that much money in chickenfeed!' was Tom's conclusion.

Fiona and Tom were not alone in their transatlantic discussions. Les Fyfe and Joe Simpson were also burning up the phone lines between England and California.

'We can't wait till October, Joe. We need to start scaling up your facility now. Otherwise we'll never have enough ready for the launch of 'Let them eat cake', in March'.

'Can't you postpone it by a couple of months? What's the rush. We own the patents'.

'No, we can't ruddy well postpone it by a couple of months. I say we can't do it. Archie Ramsbottom's spending a fortune converting one of his plants to produce the capsules. They've just about worked out the final formula and he says they'll be ready to go into full production by the end of the year. We need to have a good stock of Compound X by then to feed into this process'.

'Les, you know it's a complicated thing to make. We're talking a seven-reaction process, and we've no idea yet if the scale-up we have planned will work'.

'All the more reason to get on with it, then. The quicker you start, the quicker you'll iron out any problems'.

'But …'

'Don't 'but' me laddie. I say don't 'but' me. You've been given every facility you could wish for. I've persuaded the Todd Board to divert hundreds of thousands of dollars into this project. Just you get on with it'.

'Les. We have to reconstruct an entire laboratory and bring in a lot of new equipment for this scale-up. Tom Carroll's bound to get suspicious'.

'Which is why you should have made sure he's in Africa well before October. Can't you find some way of getting him out of the Todd between now and then? C'mon, you are his boss'.

'Well, he did say he'll be travelling to some conferences over the summer. That's why he couldn't leave for Africa sooner'.

'There you are, then. Just plan the reconstruction for his absences'.

'Easier said than done. This is a big deal'.

'Let me tell you about a bigger deal, Joe. Just imagine an evening in March next year on board the Royal Yacht Britannia, specially booked for the launch of 'Let Them Eat Cake', with a Royal Highness thrown in and the cream of the London media on board. All paid for by Archie Ramsbottom. Only, nothing to launch. Oh dear, our California facility couldn't deliver the goods in time. Sorry, Mrs Ramsbottom, you've squeezed your size fourteen figure into a size twelve designer frock in vain and have cut off large parts of your circulatory system for nothing. Laddie, that's what I call a Big Deal, and it's not one I'm prepared to contemplate. Get that facility up an running!'

'OK, OK! We'll make sure it happens'. Joe Simpson knew that there was no point in arguing with Les when he was in this kind of mood.

Since his return from Cap Ferrat, Henri de Klompenmaker had also been engaged in extensive discussions.

'You've seen The Chairman's memo, then?' were Miel's first words as she strode into Klomp's office at 10.00 a.m. sharp.

'You too?'

'No, but I've heard about it from my contact at Head Office. Big cuts, I hear'. Miel could say this with even more than her usual insouciance, as she knew that the Management Trainees were a protected species, never to be culled. 'The future of SI' was how The Chairman had described them. Her position was secure.

Klomp knew this too, which meant that Miel was the one person in London with whom he could discuss his proposed cuts frankly.

'Shall we get down to business? Have a look at my memo'. Klomp stared out of the window while Miel read. He wanted to send his response to The Chairman by the end of business today. That would impress them at Head Office. Now that his secondment in London was in its final year, Klomp had to position himself for his next move. Should he tell Miel what he had in mind? She was sure to leak it to her contact in the Chairman's office, which might be helpful. But perhaps not yet. He would wait until his cost-cutting plan was sent. Yes, that's when he would tell her about his hopes for the Presidency of the SI Leisurewear Division. That's when The Chairman should know of his aspirations. Winning The Snurt, was a good start, but offering draconian cuts in his own department should clinch it. This would show his mettle and he would further ingratiate himself with The Chairman. His time on the flight home had been well spent.

'Well?' Klomp was eager to now what Miel thought.

'10% staff cuts across the board. That's a lot', said Miel when she had finished reading.

'We can offer even more. Here's my plan'. Klomp walked over to the flip chart and pulled back the top sheet. The figures spoke for themselves.

'Staff: 10% reduction in headcount; 15% reduction in salary costs'.

'You're getting rid of the expensive people', was Miel's perceptive comment.

Klomp nodded'.

'Travel: 15% reduction'.

'New York Department: closure'.

Miel was impressed. 'Closing an entire department. That's spectacular. They'll love it at Head Office'.

Klomp elaborated on his plan. The big saving would be in closing the New York operation. What with the fax and the telephone, there was no need to employ people there to do business in America. The office there was full of old timers who had been there for years. Redundancy payments in the US are negligible: they'll be cheap to lay off. In London, the focus would be on laying off middle managers, partly to make room for the management trainees. There would be less travel all round. Overall saving should be around 15%.

Miel knew that the plan would play well in Head Office, and that Klomp would be flavour of the month as a result. By association, she would too. So far, so good. But there were two aspects of Klomp's plan that gave her grave concern.

'Henri. What do you mean by laying off middle managers to make room for management trainees. Are you expecting us to do a job, I mean their job?'

'Correction. I expect you to re-engineer their job. Make it more efficient, process-driven'.

'Yes, I understand, but I was always told that we are to keep our hands free, not get too deeply involved in a particular function, keep a distance, be ready for the next move'.

'And my other concern is with cutting travel. We do need some travel to do our business'.

'You're right, Miel, and essential travel will be unaffected. What I want to cut down on is the unnecessary travel to meet authors and editors. All that can be done by phone and fax. Travel between SI offices, to planning meetings, working groups, etc. will not be affected'.

'That's good to know'. Said Miel, much relieved, 'but there's another thing. I don't think you can expect the management trainees to take over the jobs of middle management. That's not what we're for'.

Klomp had thought this would come up and had his answer ready. He knew better than upset the corps of management trainees, they were too well-connected. Aware that the prospect of having to do a real job for more than six months would cause them great distress, he moved quickly to reassure Miel before she could spread panic through the ranks.

'Of course not. My plan is for a core team, under your leadership of course, to do it for a few months, with the express purpose of re-engineering the function and totally rewriting the job description. You've been here almost six months. You've seen everything that goes on. You'll know the job backwards by now. I'm confident that you could complete this new assignment by the end of the year. Look on it as a Project, not a Job. And then …'.

To the ambitious management trainee, the 'And then …' was the whole point. What would the next step be on the ladder? Miel could not resist the bait.

'And then?'

'And then, I'll be moving on. My secondment here will be complete. Thank God! If we make the right impression on The Chairman now; help him out of this financial hole, there will be opportunities'.

'Opportunities?'

'If we can show how much we can cut costs here, who knows?'

Miel's lips moistened. 'Do you want any help on your memo to The Chairman?'

'I need to get more data on salary and travel costs. Could

you organize that for me?

'For New York'.

'Yes, and for the middle managers here'.

'Names?'

'He listed five names'.

'Fiona Hamilton? Isn't she meant to be quite good?'

'A necessary sacrifice. Could you have the information by 2.00 p.m. That will give us a couple of hours to work on the text'.

Fiona was summoned to an audience with Henri de Klompenmaker the following week. This was, she assumed, the meeting that she herself had been seeking, to discuss her California trip.

'I haven't read your Trip Report. Too busy. I wanted to see you about something else. We're having a reorganization'.

'Another one?'

'Don't you read the financial press?'

'Of course I do'.

'Then you'll know about the fall in the price of SI shares'.

'How could I not, we all keep an eye on our stock options. Sounds like hanky panky in the agricultural machinery division'.

'The Chairman wants the rest of us to make up the shortfall in profits'.

'That's not going to be easy. The gap must be huge'.

'Well, we're going to do our bit. I've submitted a cost-cutting plan to The Chairman, and he's accepted it. We're going to have to make some reductions in personnel and re-engineer our processes to make them more efficient'.

Fiona caught his drift.

'Do I take it that I'm to be re-engineered?'

'You could put it that way'.

His smirk as he said this was enough to reveal his plan for her.

'In other words, I'm out?'

'Clever girl'. Klomp waited for the tears to start. He had been here many times before during his two years in the London office.

'Don't call me 'girl'. I don't agree with this at all. My department is doing very well. Rapid growth and record profits. You've got no grounds for making me redundant'.

'You've got no choice'. Klomp had not expected so robust a response. As he was to have five of these discussions this afternoon, he had ordered a box of paper handkerchiefs and had two HR executives on standby.

'We'll see about that. I know my rights and I refuse to continue this discussion without a witness'.

'A witness? I can call in somebody from HR'.

'No way. I'm not having one of those bimbos. I want a witness of my own choosing. And, unless you want to discuss my report on the California trip, I consider this meeting to be at an end'.

'If that's the way you want it. Will that be all?' The tone and pace of their discussion had taken Klomp by surprise. No tears at all.

'Not quite. I want your redundancy proposal in writing? I'm going to take advice'.

'Advice?'

'Advice with a capital 'A'. I'll be talking to my lawyer'.

'Lawyer. I think you are over-reacting, don't you. I'm sure we can deal with this internally'.

'You must be joking, I'm reacting exactly in accordance with that employment law workshop you insisted on all the managers going to last year. It wasn't a complete waste of time after all. I want a proposal in writing, including a detailed calculation of my severance payment'.

Klomp had forgotten about that course. He should have been better prepared, and did not like the idea of lawyers becoming involved. 'OK. I'll arrange it'.

'When?'

'I'll try for the end of the week'.

'Not good enough, Henri. I want it tomorrow and I'll set up an appointment with my lawyer for the day after'. With that, she swept from the room without further formality, snorted at the attendant HR executives waiting to counsel her, and headed straight for her office. Once there, with the door closed, her indignation was vented fully.

'How dare he! That book-keeper! That ignoramus! Me, fired by that? With my programme doing so well. I'll fight this'.

She paced up and down in her room as she marshalled her thoughts. She did not want to speak to anyone just yet and told her secretary to hold her calls.

As her anger cooled, Fiona began to think more clearly. She did not actually like working for SI under the Archer regime and held her boss in utter contempt. In many ways, the organization was becoming an embarrassment to represent. Standards had fallen and were bound to fall further. She knew she was highly regarded in the industry. Head hunters had approached her twice in the last six months alone. Getting another job would not be a problem. The redundancy payment should be substantial. And there were her share options. She would be able to vest them immediately. The share price was high now, but what about the future? With this management it was bound to drop. Was redundancy now actually an opportunity in disguise? The words of her first boss came back to her. 'There's only one thing worse than being fired, and that's not being fired'. He had certainly done very well out of the last major re-engineering. What with his payoff and the cash from his stock options, he had set up his own business and was now doing very nicely, thank you. She sat at her desk, laid one sheet of writing paper upon it, and wrote a list of pros and cons of redundancy. Thirty minutes of deliberation had resulted in only three 'cons' – loss of company car, pension and five weeks paid holiday, while she had stopped writing down the 'pros' after the first ten. The pros definitely had it.

'I must call Cousin Kirsty'. Cousin Kirsty happened to be one of London's brightest young employment lawyers.

Before she could do so, her phone rang.

'Maria, I said no calls'.

'It's Michelle from HR. She's here and says she wants to counsel you'.

' What? Please tell her that she has no need to be concerned, and that I don't need counselling. Get rid of her quick and step into my room. I want to tell you what's going on'.

Maria, who had been with SI since leaving school, was initially tearful. She liked working for Fiona and did not relish the prospect of another boss. But she began to cheer up when she realized that she too might have the opportunity to be re-engineered out of the company.

'I can't promise anything, but I'm sure the redundancy terms will be pretty good. They usually are in these situations. Think about it and tell me what you want to do. You've always said you really wanted to start up your own jewellery business, working from home. This could be your chance. It'll give you the money you need to start up'.

By the end of the next day, Fiona had received a letter

outlining her redundancy terms. 'Not quite good enough', was her initial reaction. 'Not nearly good enough!' opined Cousin Kirsty, who took over negotiations on Fiona's behalf. By the end of the following day, Fiona's severance package had doubled. A most satisfactory outcome. Cousin Kirsty then took up arms on behalf of Maria.

'We're not going to leave much for anyone else', said Fiona to Maria at their celebratory dinner.

Henri de Klompenmaker was not in such a jovial mood. In his daily briefings with Miel Flick he kept her abreast of developments.

'She's going to cost us a fortune. And her secretary too'.

'Yes, but it doesn't come off your budget, does it?' Was Miel's instant response. 'It will all come out of the 'Reorganization' provision at Head Office'.

'Yes, but why should she get so much? It's not fair. This lawyer of hers has squeezed a lot more out of the HR people. They say we have no option. Did you know that she has to be allowed to cash in her stock options now? The share price has bounced back since The Chairman's statement on cost savings'.

'Thanks to you, Henri'.

'Yes, that's what sticks in the throat. I can't cash in my options for another year'.

'The other thing is, she wants to leave at the end of next week. Turns out that she's taking all the unused holiday she is due. It amounts to almost three weeks. We offered to pay her for them, but she doesn't want to stay until the end of the month'.

'Very unprofessional'.

'That's what I told her. Of course, it means that you'll have to take over her programme rather sooner'.

'Me? Take over? Hold on there. I though t I was going to be re-engineering the processes. Kind of like an internal consultant. Treat it like a project, not a job, you said'.

'Yes, exactly. Not a job, but on the job'.

At this moment a *froideur* entered into the working relationship between Miel Flick and Henri de Klompenmaker and the chill would intensify in the coming weeks.

Fiona Hamilton, meanwhile, was warming to her topic in her telephone call to Tom Carroll.

'The cheek of that man. He told me I was being unprofessional, because, having been made redundant against my wishes – at least to begin with – I now want to use up my full holiday entitlement. Unprofessional, *moi*? Look at him. Apart from the mess he's making here, he's in the process of destroying our entire New York office, and is putting our most valuable publications in the hands of management trainees who are barely out of nappies. I tell you, Tom, it's Captain Smith on the Titanic going full steam ahead for the iceberg'.

This was Fiona's second phone call to Tom in as many days. In the first, he had persuaded her that her redundancy was indeed an opportunity, not a threat. He pointed out to her that she was obviously unhappy with the direction the company was taking, that her severance package was beyond the wildest dreams of any American, and that, if she was correct about the Titanic scenario, they were doing her a favour by throwing her into a lifeboat now.

As a sign of his displeasure at her inconveniently imminent departure, Klomp allowed Fiona only one trip to bid farewell personally to a journal editor; Les Fyfe. The rest she could do by phone. He had wanted Miel Flick to accompany her, but Miel seemed not to be as enthusiastic to assume her new responsibilities as he had hoped.

'Rochdale? You must be joking! I thought you said that one of my objectives was to cut back on the unnecessary travel. I already met Les Fyfe during one of his visits here, if you recall'.

Although he did not like Miel's newly less-than-co-operative manner, Klomp did not press the point. He knew that Fiona would get up to no mischief with Les. She knew that her severance cheque would be threatened if she did not use her 'best efforts' to ensure a smooth transfer of her responsibilities. As a precaution, Klomp called Les Fyfe to inform him of Fiona's impending departure. Les's joy was unconfined. That evening he and Vera celebrated with a bottle of Krug, courtesy of Bear State University. They agreed that it would be great sport for Les to phone Fiona, pretending to be ignorant of her plight.

'I say, Fiona, is that you?'

'Yes, Les'.

'I hear you want to come and visit us next week, do you? It's a bit short notice. What's it all about? We'll be wanting to make preparations'.

'Actually, Les, I wanted to let you know that I'll be leaving SI, and we've got a few loose ends to tie up on the journal'.

Despite his prior warning from Klomp, and his unusually sunny demeanour since, Les Fyfe managed to feign both surprise and distress at Fiona's news.

After his expressions of thanks and sadness, as insincere as they were verbose, he got down to business.

'You'll be putting together an Agenda for our meeting, I take it?'

'Of course'.

'And you'll be wanting us to book somewhere for lunch?'

'Dinner, actually'. Fiona wanted an excuse to stay overnight in the area.

'I've so much to do here, I'll have to come up in the afternoon'.

'Dinner it is, then'. Les was delighted; when eating at SI's expense, he much preferred the more lavish evening entertainment.

'Will you want us to be booking you a hotel room?'

'Don't worry, I'll take care of that myself'.

When Fiona put down the phone, she called for Maria, her secretary, and together they pored over the hotel guides for Lancashire.

'I want somewhere near Oldham, Maria'.

'Oldham? But I thought that Professor Fyfe was in Rochdale?'

'He is, but I want to stay somewhere a bit different this time'.

'What about this place. Chadderford Hall Country House Hotel. 'Rural charm with good connections'.'

'Looks nice, but it's not cheap, is it'.

'Travel in style, for once, Fiona!'

'Oh, why not. It will make up for Tallinn. Book me in!'

The following Wednesday morning, Fiona's little BMW was heading north towards Lancashire. She wanted to get there by mid-day, so that she would have time to check out the Pennine facilities in the area. Her inquiries had revealed three. Two production facilities (one in Burnley and one in Oldham) and one research facility, in Rochdale, which also served as the company headquarters.

She headed for Rochdale first and waited in her car outside the Pennine building for the best part of an hour. People came and people went, but nothing unusual happened. As the afternoon was wearing on, she decided to drive the short distance to Oldham, finding the Pennine facilities there only with great difficulty. Unlike the other plants, which declared their existence with lavishly belching chimneys and large, rooftop billboards, the Oldham plant had only one obvious sign.

'Closed! What a waste of time'. Then she saw the vehicles in the car park.

Fiona was peering through her binoculars at three cars and two trucks parked against the factory wall, when a loud honk from immediately behind made her jump'.

'You don't want to be standing there, lass, you'll get run over!' Fiona turned round and saw the portly figure of a driver as he jumped down from a large truck.

'You waiting for somebody?'

'Oh, no, no, I'm a, er, a birdwatcher. Yes a birdwatcher. Look, you've frightened him off'.

'Sorry about that, lass, but I've got a delivery to make here. I'll need to open gates. Can you stand back?'

'Looks like a big load'.

'Big, you're telling me. They're re-engineering a whole plant in there. Some new product line they want to get going'.

'Oh, that's good news. More jobs for the area'.

'If the Yanks leave us any, that is. The place is swarming with Americans'.

'Gosh, that's exotic'.

'You're telling me. From some university in California. What they want with Oldham, I've no idea'.

Fiona could barely contain her excitement. 'A lot of them, are there?'

'Well, when I say swarming, I mean four or five, but that's a lot of Yanks for these parts'. He finished pushing the gates open.

'Do you know what they're doing?'

'Are you birdwatchers always this nosey? Sorry, lass, I've got a job to do and I can't stand around gassing all day'. He climbed back into his cabin, pulled forward into the yard and, Fiona thought, closed and locked the gate in a very pointed manner. 'Happy hunting'. Were his parting words.

Fiona hastened to her hotel and called Tom as soon as she checked in. He was attending a conference in Boston and she was lucky to catch him in his room.

'Joe's got such a big group, I never know who's there and who's not. Remember, I'm not there much myself at the moment. This is peak conference season. If you could get me some descriptions, I might be able to get Susan to identify them, but they're almost all oriental, so that might not be much help'.

'Couldn't you ask around?'

'Difficult. Joe's group hardly talk to me and if I start asking questions, he'll only get suspicious. We don't want to blow our cover at this stage'.

'True. I'll go back to the factory tomorrow, and try to get a sighting. I've got my binoculars'.

'You'll need them. They all look the same'.

'Tom! You're not meant to say that sort of thing nowadays'.

'I can to you, Ms WASP'.

'Look here, I've got to get ready for my dinner with Les. I told him I'd meet him at the restaurant'.

'Good luck. Make the most of it'.

Fiona drove herself in her little BMW to the 'Restaurant Versailles' on the outskirts of Rochdale. Having allowed extra time to get lost, she didn't, and arrived a full fifteen minutes before the appointed time. The concierge at Chadderford Hall had told her that she could not miss it, and he was right. The light shining through its stained glass windows illuminated the entire street.

As far as she could tell, she was the only guest in the restaurant that evening and she received the full, undivided attention of Gaston, the Maitre d'.

'Would *madame* like to be shown to the table directly, or would you prefer to have a drink first in the Pompadour Lounge?'

'I'll have a drink in the lounge, thank you'. She was shown into a room decorated in a repellent blue, with Empire-style furniture and a gothic bar which occupied an entire wall. It looked like a shrine and she could hardly take her eyes off it.

'I see madame is admiring our bar. It is unusual, *n'est ce pas*? The wood panelling was stripped from an old chapel before it was torn down. You will find it throughout the restaurant. The altar screen is used here, for the bar. The communion table is in the Salle Louis XIV, our main dining room, and you might have noticed the baptismal font in the hall. It makes a wonderful setting for flower arrangements'.

'Yes, most unusual. But, if I recall correctly, Versailles is not particularly gothic. I don't understand...'.

'*Vraiement, madame, vraiement,* Versailles is what we call baroque. We originally named the restaurant 'Chartres' – *tres gothic, n'est ce pas* – it is my home town, but nobody here had heard of it, so we changed to something more recognizable'.

'That makes sense. Chartres. Hence the stained glass windows'.

'*Exactement*. It gives such a beautiful light during the day. What would *Madame* like to drink while she waits?'

'I'll have a tonic water, I'm driving'.

Sipping her drink, Fiona examined the pretty, engraved glass and saw that there was writing around the rim. *'Qu'ils mangent de la brioche'*. There it was also on the bowl containing the nuts. And on the napkin. 'Must be their motto. *Tres amusant,*' she thought.

Gaston returned to ask if she needed anything else.

'Very clever motto for a restaurant, Gaston. You're idea?'

'Bien sur. Vous parlez francais, Madame?'

'Oui, j'ai passé beaucoup de vacances en France. Le francais est ma langue preferee'.

'You are one of the few to get our little joke. So few English people speak any French'.

'Monsieur! Je suis ecossaise!'

'Ah. *Mes excuses, Madame. J'aime bien les ecossais.*

Fiona passed the ensuing ten minutes in conversation with Gaston, discovered that he had been living in England for 15 years, that he moved here for reasons of the heart, and that Professor and Mrs Fyfe were among his most loyal customers. As a special honour he allowed them to bring their own wine.

'They have such a wonderful wine cellar, it is an honour to open their bottles. Only last week they arrived with a magnum of Chateau Latour 1961. I ask myself', How does this man in Rochdale get his hand on such fine vintages?' He is a marvel, Professor Fyfe. *N'est ce pas*? They say he will win the Nobel Prize one day'.

Fiona, true to her upbringing, managed to control herself in the face of this startling news. The tonic water she held trembled only slightly. She was sorely tempted to reveal to Gaston the source of Les Fyfe's fine wines, but was saved from doing so by the distant sound of an all too familiar voice. Les had arrived and was half way into the

Salle Pompadour before Gaston could intercept him.

'*Monsieur le prof.* So nice to see you again'.

'Thank you Gaston. Where's Dean tonight? There's nobody in the hall to take coats'.

'As we are quiet, he has gone to visit 'is mother in Bury. I take care of the front-of-house myself. My mother is in Chartres. I never see *maman*, he always tells me that I cannot be away for so long. He sees his *maman* every week, but I never see mine. Is this fair, *Monsieur le professeur*?'

'Who said life was fair, eh, Gaston. Let me have a look at the wine list, will you? I fancy some champagne'.

'You did not bring any bottle this evening, monsieur? I always enjoy the surprise. You have such wonderful wines'.

'No, not this evening. Hello Fiona. I take it that SI will be paying this evening'.

'*Comme d'abitude, Les*'.

'I'll take that for a 'yes'. Gaston, the wine list, if you please'.

While he studied the wine list, and subsequently the menu, Les attempted to convey to Fiona a sense of sadness and regret at her impending departure. He failed to do so, and Fiona briskly curtailed his efforts by turning the discussion to champagne and urging him to choose the Dom Perignon.

He readily did so, and had already consumed three glasses while Fiona sipped occasionally from her first. When Gaston announced that their table was ready Les instructed him to keep the remainder of the champagne on ice to accompany their dessert.

The Salle Napoleon was as austerely gothic as the Salle Pompadour. The only diners that evening, Les and Fiona sat in a booth constructed from church pews, rendered only barely comfortable by a liberal scattering of cushions. They completed the journal business before they started the main course, at which point, Les called for a second bottle of red wine. Things were going the way Fiona wanted. She judged him to be sufficiently relaxed to move the conversation into more treacherous waters.

'Tell me, Les, what are your plans for the future. You've been editing *Transactions* for a long time'.

'Funny you should say that, Fiona. I've been giving my future a lot of thought lately. I can speak to you as a friend now, can't I? You'll not be carrying tales back to Henri de Klompenmaker, will you?'

'Of course not, Les. My lips are sealed'.

'Well, I've got more strings to my bow than *Transactions*, I can tell you'.

'I know that. Your Todd responsibilities must keep you quite busy'.

'It does, of course, but I'm not talking about the Todd, I'm talking about private enterprise. I'm to be rich'.

'Well, I'm very happy for you. What is this private enterprise?'

'I can't say anything more about that now, but suffice it to say that this time next year SI might be looking for another Editor'.

'Gosh. I can hardly imagine *Transactions* with out you. That will be a terrible shock to the community'. Fiona, with great effort, managed to make these utterances with a straight face.

'Very kind of you to say so, Fiona, very kind, but they'll just have to manage. They won't have Fyfe to kick around any more. But don't let on to Henri. I want to tell him myself. I'm also hoping to take early retirement from the university'.

'Early retirement? This must be quite an enterprise'.

'It is, lass, it is. All I'll say is be sure to keep a close eye on the financial pages early next year'.

'Is it a company you've invested in?'

'In a manner of speaking, but I can say no more now'. He changed the subject. 'Pity you can't drink a bit more of this wine. It's luscious. I say it's luscious. Why didn't you get taxi?

'Well, I'm staying a bit out of town. I prefer to sleep in the country if I can'.

'Oh, where are you, then?'

'It's what they call a country house hotel. Chadderford Hall'.

'Les almost dropped his glass'.

'Chadderford Hall you say, what the hell are you doing there?'

'I beg your pardon! It's a free country, isn't it?'

Les realized he had over-reacted. 'Oh, I don't mean that. It's just, it's a very pricey place. I'm surprised SI let you stay there'.

'It makes up for Tallinn'.

'Yes, I suppose so. You there just the one night, then?'

'Yes, unfortunately'.

'Busy, is it?'

'The desk said they've been very busy lately. Masses of overseas visitors apparently'.

Les blanched.

'Overseas? They shouldn't be giving out details about guests. It's not allowed'.

'No details, just 'overseas'. Actually, I've not seen anybody, from here or abroad'. To Fiona's disgust, Les seemed to have become alarmingly sober, alert almost.

'Well just you keep it that way. You know what

foreigners can be like'.

'Les, you can't say things like that'.

'Would you like me to come back with you?'

'Les!'

'Oh, nothing like that, just to make sure you're all right. It's dark now, and you might lose your way. I can get a taxi home from there. It's practically on my way, any road. I insist. I won't be able to sleep until I know you're safely back there'.

'Very well, but you behave yourself, mind. Dessert?

The rest of the meal passed in virtual silence. Les and Fiona were absorbed in their own thoughts. Both, as it happened, on the same topic.

'I've got to make sure she doesn't find out about the Todd people in the hotel. I'll wait and check that she goes straight up to her room, and then tell them to keep a low profile'.

'Why is he so agitated about my hotel? What's got him so worked up about the foreign guests. I wonder …'.

'I expect you'll be wanting an early night, Fiona. You've had a long drive today'.

Fiona thought her best policy was to break the habit of a lifetime; she agreed with Les.

'You're quite right, Les. I need an early night. Shall we have our coffee and go'.

Les could not believe his ears. She had never agreed with him without a fight. Something was definitely afoot. He would have to be on his guard.

As Fiona drove along the dark, country roads she could see the surrounding hills bathed in a full moon. Les was keeping very quiet and she felt that a conversational stimulus was called for.

'Aren't the Pennines lovely in the moonlight'.

The sudden movement of his considerable bulk caused the car to rock slightly.

'Eh, what's that about Pennine?'

'I said, 'Don't the Pennines look lovely in this moonlight. You know, the hills!'

'The Pennines? Oh yes, fine hills, very fine hills. I say, we can't be very far from the hotel now. I'll tell you what. Why don't you drop me off at the front door before you park your car, and I'll get them to phone a taxi'.

'Yes, of course, Les. It's been very kind of you to escort me back. Will you stay for a drink before you head home? They've got a lovely bar, and a fine selection of malt whiskies, I believe'. Les Fyfe had never been known to turn down a free drink, and he liked to think himself a connoisseur of Scotch malts. His response was immediate and entirely out of character.

'Sorry, Fiona, I'll have to decline. I can't linger tonight'.

'But you'll be waiting for your taxi anyway?'

'Er, oh, well, I mean, I wouldn't dream of imposing on you. You should get straight to bed. You look dead tired, lass'.

Fiona had rarely felt friskier, but she humoured him. 'Oh dear, do you really think so? I suppose I have had a very busy week. You're probably right. An early night would do me good'.

This was the second time in her relationship with Les that she had agreed with what he said; both times this evening. Perhaps she just wanted to end their working relationship on a positive note.

'At last, here we are'.

Fiona's little BMW crunched along the gravel drive. Les had already unfastened his seatbelt and leapt out of the car as soon as she came to a halt. She had never seen him move so quickly. Parking just beyond the door, she locked up and in an instant was on the threshold of Chadderford Hall. By then, Les was already trotting into the bar at the far end of the hall. Two front desk clerks and a bell boy were staring after him. Fiona moderated her own pace to a brisk walk. Officially, she was making for the staircase and bed, but having bade them a good night, she made a last minute detour to the bar door, just in time to see Les accost four dark heads that were sitting in armchairs before the fire.

'Never mind that. You'll have to move, and quick!' All four rose and were ushered by Les into the adjacent conservatory. The door banged shut.

'So it is them!' Fiona almost shouted. As she would hear no more now, she casually strolled back towards the front desk. Les did not linger in the Conservatory.

'Sudden call of Nature. My age, I'm afraid'.

Fiona smiled sympathetically. The desk clerks looked surprised. The toilets were in the other direction.

'You get off to bed, now, and I'll sort out my taxi'.

'It's been a pleasure working with you. Good luck, Fiona'.

'Thank you, Les'. They shook hands and Les Fyfe watched her closely as she mounted the stairs. She waved.

'Hm. He'll have warned them about me, so I've little hope of getting anywhere. Still, nothing ventured ...'. She would lie low in her room until she heard Les's taxi depart.

Fifteen minutes later, a crunch of tyres on gravel indicated that a vehicle had arrived. Fiona watched as Les stepped into the taxi, and then skipped downstairs to the bar.

'Empty!' She sidled over to the conservatory door and peeked in. 'Nobody! Unless they were hiding among the potted plants'. Before exploring further, she decided that she would look more convincing with a drink. She rang for service.

'A Glenmorangie, please, no ice'.

Armed with her drink she felt it was safe to drift casually into the conservatory. To her chagrin, no foreigners lurked in the foliage. Disappointed, she decided that she might as well have a seat by the fire and enjoy her whisky. She began to make plans for her return visit to the Oldham plant tomorrow.

As she gazed into the glowing embers, she heard a distinctly foreign voice.

'Hey, you guys. You should've sent somebody up to rescue me. Joe's been on the phone for over an hour. I need a drink. Guys?'

Evidently there was a fifth. Fiona looked round from the depths of her winged, leather armchair.

'Excuse me, ma'am, I thought my colleagues were down here'.

'Sorry to disappoint. I'm here on my own. Why don't you join me for a drink? Sounds like you need one'.

'Thank you, ma'am, I'll be happy to. Dr Chai Ling, pleased to meet you'.

'Cordelia Young'. Fiona decided that some camouflage would be prudent.

'Not many guests in the hotel'.

'No, I've only met one other guest. She was down here earlier. Fiona Hamilton, if I remember correctly. Have you met her?'

'Fiona Hamilton? No, though the name is familiar'.

Dr Ling rang for a drink. Fiona, aka Cordelia, declined his offer of a second.

'You don't sound English. Have you come very far?'

'Actually, I'm Taiwanese, but now live in California'.

'How do you like it there?'

'I like it very much. Californians are very friendly'.

'So I've heard. What brings you to England?'

'We, my colleagues and I, are working on a project here'.

'It's an out of the way place for someone from California'.

'I can't really talk about it. Sorry'.

'Oh, do please forgive me. I shouldn't be so nosey. I expect you'll find the weather a bit of a shock'.

'Yes, so much rain, and so cold! But we're working, so we don't notice it that much'.

A silence descended.

'I expect you'll be looking forward to going home'.

'Yes, we go back next week, when the other team comes over'.

'Such a pity you don't have a chance to see more of England'.

'I'll be back again. Our project goes on at least until the end of the year'.

'I've been to California once. We went to the Napa Valley. Had a wonderful time at a winery. I expect you can do that every weekend'.

'Actually, we live in Southern California'.

'Really, but at least the ocean's warm. I expect you live near the coast'.

'Right on the coast, near San Diego'.

'I hear the climate there is wonderful. Sunshine all year'.

'Almost. Nice temperature. No humidity. Very different from Taiwan'.

'Have you been in California long?'

'Yes, quite a few years. I studied there for a PhD in chemistry'.

'Really! A good friend of mine did research at the University of California. She loved it'.

'Actually, I am at a private university. You probably do not know it. Bear State'.

'Bear State! Of course I know it. I'm a very keen windsurfer. It's the Mecca'.

'Windsurfing. Very amusing. I did not think it would be a big deal in England. So cold!'

'Well, you do have to be keen. But I am. I've read a lot about the Todd. It's an amazing facility'.

'Actually, that is where I am based'.

'Really, you a windsurfer too? How thrilling!'

'No, there is also a Todd Research Institute. That is where I am. What brings you here, ma'am?'

'Oh, bird watching, it's a hobby of mine. Lots of interesting birds in the Pennine'.

'Pennine? I didn't know you had a connection'.

'A connection? Yes, and my father too'.

'With Pennine? Do you know Mr Ramsbottom?'

'Which one? There are lots of Ramsbottoms round here'.

'The Chairman of Pennine'.

'Pennine? I'm talking about the hills. The Pennine Hills. They're all round us' Fiona thought it best not to show too much interest in the company. She'd established the connection.

Having realized his *faux pas*, her drinking companion had less to say from then on.

The conversation continued, haltingly, until their glasses were drained. Fiona did not want to show too much interest in his work, nor did she want to dwell on windsurfing, where the slightest probing from her companion would reveal her ignorance of the sport. As he made little effort to initiate any topic, they eventually lapsed into silence. Dr Ling, as soon as he could decently do so, rose from his chair and bade her good night.

Fiona reviewed her campaign to extract information from him and concluded it had not been a success. 'Nothing I didn't suspect already; but it must be a big project, if they have teams here until the end of the year'. She was very pleased with her cleverness in assuming an alias, and particularly liked the bird-watching twist. *Carte blanche* for binoculars. 'Les will have warned them off Fiona. Cordelia will have to stay in her room until they leave tomorrow morning, though. Breakfast in her room and a nice long bath, I think. I must phone Tom about Dr Ling'.

The next morning Fiona awoke early. It was raining heavily. From her window she saw that there were two cars and a minivan parked out front. She ordered breakfast for eight, made herself a cup of tea and reviewed the plans that she had made the previous evening. Once the Bear contingent departed she would check out and drive over to Oldham. There she would test the defences of the Pennine plant. The fencing had looked rather decayed and she was confident that she would be able to squeeze through somewhere. Her ornithological alias would explain her anorak and binoculars; useful cover if caught. None of them had seen Fiona Hamilton; if she happened to bump into them, Dr Ling would introduce her as Cordelia Young, a birdwatcher. Yes, all pretty convincing.

Later that day, driving south and soaked to the skin, Fiona was malodorous, but triumphant. Things had not gone exactly according to plan. Having squeezed through a gap in the fence, she had made it to the wall of the factory itself.

To her disgust, the windows were too high to see through. She tried various doors: all but one were locked and this gave access to only to a small storehouse. In there she found nothing more than stacks of empty jars and piles of paper labels. In the gloom of the store she could not read the labels, but she stuffed one into her pocket and used others to wipe the mud from her boots. She continued her probing round the factory wall and eventually came upon an old barrel, which, if mounted, would allow her to see through a window. Having climbed onto this, she peered inside. There they were. Five orientals and two white men, putting together some kind of enormous chemical apparatus. There were vats and pipes all over the place. This was something! In her excitement she had not heard the first crack in the rotten wood of the barrel top. She felt the second, but by then it was too late and she plunged up to her waist into a stinking, dank slime. She had the presence of mind not to scream, but fearing that her fall had been heard inside, pulled herself from the barrel and headed back to the gap in the fence. Just as well. They had heard, and two men were running toward the barrel. They looked into it, then looked around. She lay low in the undergrowth by the fence. The wet, muddy ground had covered her tracks and they did not seem inclined to linger in the rain. Nor did Fiona, and as soon as they disappeared back into the factory, she made for her car and was off; and not a moment too soon. As she drove past the entrance gate, the Mercedes she had seen at the hotel emerged and sped up the road she had just driven along. Her heart was in her mouth. She pressed the accelerator to the floor and did not let up until she was well along the motorway back to London.

'It's industrial espionage, I tell you. Industrial espionage'. Archie Ramsbottom had arrived after lunch and was standing in the rain beside the fractured barrel. 'We need to increase security here. More bloody expense. Have you told Les Fyfe?'

They had, and he was on his way. At an impromptu Board Meeting of Pennine that afternoon, it was decided that the Oldham Plant windows should be blacked out and 24-hour security guards appointed. 'I reckon this Thin-Kwik's going to bankrupt us before it makes us a fortune'. Was Archie's considered opinion at the end of the proceedings.

By that time Fiona had reached home and was soaking in a bath scented with aromatic oils. Her clothes, which were both dry and distinctly ripe well before London, had been thrown into the washing machine upon arrival. Refreshed by her bath, she called Tom.

'They're planning to produce something there. I think it must involve Compound X'.

'But what could it be?'

'Animal Supplements. That's what Pennine specialises in. It would fit in with Prunella Todd's weight-loss experiments on Missy'.

'Thin dogs? You're joking. Even Les couldn't imagine that a doggie diet aid is going to make his fortune. I mean, if you don't want a fat dog, just feed it less. They're not like cats, they don't hunt. Compound X must have some other effect that they're not letting on about. It could be behavioural. They claim that this stuff does act on the brain, after all. Maybe it has other effects. What do you think?'

Silence at the other end of the phone.

'Fiona, are you still there?'

'Of course I am!'

'Well, what do you think? Other behavioural effects? Could it make them less aggressive? More docile? Then they would put on weight. Yes, that's more like it. Fiona, are you listening to anything I'm saying?'

Fiona stood up suddenly. 'Just a minute! I did pick up a piece of paper at the factory. It was like a label. O My God! I put it in the washing machine. Hold on!'

Fiona extracted her washing from the machine. She was not hopeful; the cycle had been 'heavy soiled'. Rummaging through her anorak pockets she found a mushy lump that she thought had been a label. She took it through to the living room and carefully unfolded it. Most of the ink had run, but she could make out the remnants of two words in large black print. She read them out to Tom: 'Thin-Kwik'.

'I told you so'.

Chapter 13: A Dinner in Hall

Sir Henry and Lady Wiseman had returned to Oxford from their summer home in Antibes. This, along with the other little 'extras' that they both enjoyed so much was funded by Sir Henry's various non-executive directorships, which had proved a generous and reliable source of income for many years. Alberta, Lady Wiseman, was busy preparing the Master's Lodging to receive their old friends, the Jacobsons, for a stay of unspecified length. Her husband, with his customary vagueness on domestic matters, had declined to say how long Sol and Elvira Jacobson would be with them. He and Sol had an important paper to write and he would be in Oxford until they wrote it. That was all very well, but his wife was expected to entertain Elvira Jacobson and she was a very difficult woman. Bertie Wiseman found most other women difficult. She was not a woman's woman. By the third day of the visit, with her husband and Sol Jacobson closeted in the Master's library, she had felt obliged to dispatch Elvira, who was always in search of a cause, to stay with her old friend Drusilla Snape, MP, who was in the throes of one of her periodic enthusiasms for righting injustice in Africa. Lady Wiseman rightly thought that this would keep Elvira occupied for at least a few days.

In the event, it kept her busy, as well as entertained for a full two weeks, during which she accompanied the MP to Blackpool for the Labour Party Conference.

While his wife was on tour, Sol Jacobson wrote his paper with Sir Henry Wiseman. Entitled '*A Proposed Molecular Basis for Dementia*', they both felt it was one of the most significant papers either of them had written. This gave the two septuagenarian Nobel Laureates a tremendous thrill. They could still show those youngsters a thing or two. It had all stemmed from that ill-judged presentation of Les Fyfe's in Tallinn. As had been amply demonstrated on the occasion, they both thought his conclusions were deeply flawed. Their subsequent discussions had resulted in a plausible new reaction mechanism that might have implications for dementia. Over the last few months, Sol's student had conducted experiments that strongly supported this mechanism, and categorically refuted that proposed by Fyfe. This was a paper they had both enjoyed writing.

On the day their paper was submitted for publication to *Acta Moleculetica*, Sir Henry hosted a dinner in Hall. Tom Carroll and Fiona were among the guests. Lady Wiseman had insisted that they send Fiona a message of congratulations on her departure from SI, and Sir Henry had put in a good word with his contacts at Albany Press, the publishers of *Acta*. He understood that she had been invited for an interview later that month. It seemed that there would be several causes for celebration this evening, and he had ordered the finest of the College wines to be served at High Table.

Tom Carroll and Fiona Hamilton arrived together at the Master's Lodging, where they were shown into the drawing room by Ashe, the butler. Accepting the champagne offered, they surveyed the scene before them. The room was full, and few of the faces were familiar, but Fiona's deductive powers were on fine form. 'That must be Elvira Jacobson', she said to Tom, nodding in the direction of a striking, tall, grey-haired woman. Elvira was in full flow, regaling her fellow guests with her tales of Blackpool, The Labour Party, Militant Tendencies, Fringe Meetings and Composite Motions. She had actually voted, quite improperly, on at least three of these – 'You see, everyone else had their hands up, it seemed rude not to'.

Lady Wiseman had noticed Tom and Fiona arrive and detached herself from a group of archaeologists to welcome them.

'So nice to meet you once again. We must drink to your liberation, Fiona. Ashe, more champagne for our guests'.

'You're very kind, Lady Wiseman. I must thank Sir Henry for setting up an interview with Albany Press'.

'Nonsense, it's the least he could have done. You've been very helpful to him, you know'.

'Me?'

'Yes, you got him interested in this rubbish that the awful Les Fyfe has been publishing in his journal. He and Sol have just written their paper debagging the whole thing. Henry says it's the most important thing he's written in years'.

'Wonderful! Did you hear that Tom?'

'That's fantastic. When's it to be published, Lady Wiseman?'

'You'll have to ask Henry that'. Lady Wiseman raised her voice to hunting-field volume. 'Henry, look who's here!'

The Master did not have to be called twice, and made his way across the room.

'My dear, so nice to see you again, and Tom too. Did

Bertie tell you?'

'About the paper, yes. That's very exciting'.

'Real exciting', added Tom 'what are you proposing?'

'I can't talk about it here, d'ye see, but do join Sol and myself for a nightcap after dinner and we'll tell you all about it. We submitted the paper for publication today. It'll cause quite a stir, I can tell you'.

'I bet!' added Tom.

'Tell me, Fiona, how are you getting along with Albany?'

'They've invited me for an interview at the end of the month'.

'Splendid. Do you know Marlene Pym'.

'Marlene Pym? Well, I know of her, naturally. She's a legendary figure in the business, but I've never met her.'

'Now's your chance. She's here this evening. Every so often I like to have the opportunity to berate one of the Albany directors about their lack of attention to my books. It keeps them on their toes. This year's victim is Marlene. She takes it very well, I have to say. Ashe!'

The butler, who hovered two feet behind Sir Henry, stepped forward'.

'Master?'

'I believe that Dr Hamilton and Ms Pym are placed near each other at dinner, are they not?'

'Indeed they are. Master. Her Ladyship was most particular about that point'.

'Splendid. If you'll excuse me, Fiona, I have to speak to Ashe about the procession into Hall. It's quite a trek, I'm afraid. I assume my wife warned you not to wear high heels'.

'Yes, she was very explicit about that, Sir Henry'.

'What's all that about? Nobody told me not to wear

high heels', said Tom.

'It's an English thing'. Fiona replied. She knew no more than he did about the origins of the 'no high heels' rule.

Lady Wiseman circulated among the ladies to encourage them to 'powder their noses' before they proceeded to Hall, as there would be no opportunity to do so once there. She also reminded all the guests to check the Board, on which was pinned the *place a table*.

Fiona explained *place a table* to Tom.

'Look, I'm right opposite Marlene Pym, and next to Sol Jacobson; you're between Marlene and Elvira. I think we're in for a bumpy night'.

Tom looked crushed.

The double doors of the drawing room swung open, a distant trumpet sounded and Sir Henry led his procession to dinner.

Halfway down the vaulted hall of the Master's Lodging, Sir Henry veered left and seemed to disappear into the wall, followed, singly, by his guests.

'Very Alice in Wonderland', thought Fiona, as she waited her turn to enter the hole. Glancing over her shoulder, she saw that only Tom and Lady Wiseman were left.

'After you, Lady Wiseman', offered Tom, gallantly.

'Oh no, dear, I always go last, to make sure that we don't lose anyone on the way. Look, it's your turn. Ashe will give you a hand up'.

'A hand up?' Fiona peered into the hole in the wall. She saw that a very high step lead into a spiral staircase, only one person wide.

'Thank you, Ashe. No wonder you specify 'no high heels'.

They wound their way, slowly, up the staircase, Lady Wiseman chattering behind.

'When the Queen Mother came to dine, you know she always wears hugely high heels, we had planned to go the long way round; what we call the terrestrial route. Much more dignified. But she wouldn't have it. Somebody had told her about the secret staircase and the roof, so she insisted on coming this way. Scampered up it like a five year old. She put it down to all these reels and strathspeys she's danced all her life'.

Fiona was not scampering. 'Roof?'

Her query was answered soon enough as she emerged onto a rooftop walk that skirted the cloisters. Sir Henry, preceded by a man holding a lantern, had already rounded the corner of the cloister and was making his way up a slope toward a large window of what Fiona assumed to be 'Hall'.

Tom emerged, and as he was saying 'Jeeze!' Fiona saw Sir Henry disappear into a hole in the wall beside the large window, while a trumpet fanfare sounded from within. The man with the lantern stood outside, swinging it back and forth. Fortunately, the night was clear and dry. Fiona held up her long dress and picked her way round the roof walk. Lady Wiseman's conversation did not calm her nerves. She hated heights.

'When Ronald Reagan was here last year, I thought we were going to lose him. A stupid secret serviceman had such a tight hold on him that when he slipped he almost pulled both of them over the side. And they still talk about the time Haile Selassi's Chamberlain did fall over the parapet when they were here in the 1930s. He insisted on walking backwards in front of the Emperor. Asking for trouble'.

Fortunately, she had only begun her story about David Niven breaking an ankle when they made landfall at 'Hall' and the servant with the swinging lantern helped them negotiate the threshold.

The sight that greeted them was magnificent, and merited another of Tom's 'Jeeze's'.

Above them, a beamed roof of evident antiquity disappeared into the gloom. Around them, the panelled walls were covered with portraits of college members, long-departed and striking poses that were variously sober, intense, pompous and unlikely. Beneath them, tables laden with college silver were bathed in the soft pools of light thrown out by regularly spaced candelabra, whose glows also shone upon the upturned faces of the student body of St Leonard's College, as they rose from their seats to receive their Master and his guests.

On taking his place, Sir Henry looked towards his butler, who eventually confirmed, with a nod, that all the roof-travellers had made safe passage to Hall. He then recited an incantation in Latin and sat down to the accompaniment of the third trumpet fanfare of the evening.

The 40 High Table guests fell into three distinct groups. At either end were clusters of the College Fellows. An unusually large number of them were dining in Hall that evening; word had got round that the Master had commanded wines of exceptional vintage to be served. The central dozen places were taken by the Master and his most important guests, who fell into two distinct groups. On the Master's right hand sat the 'political' group, comprising a Cabinet Minister, an Ambassador, a Peer of the Realm and a Dame of the British Empire. On his left hand sat the 'publishing' group. For the first two courses of dinner, Sir Henry devoted himself to Affairs of State with the political group, while his wife, sitting opposite, tried to give most of her attention to the publishing group: she could not resist the occasional foray into the political side of the table when she felt that they might benefit from her unique insight into Comrade Stalin's views.

Left largely to their own devices for the best part of an hour, the publishing group amused themselves.

'You must be Fiona Hamilton. Marlene Pym. *Felicitations*!'

'*Felicitations?*'

'I hear you've been fired by SI. Another victim of that ignoramus Klomp, I suppose. You're well out of it. The lunatics have taken over the asylum. What poor old Algy Brogue must make of it all, I don't know. He was always so loyal to his staff'.

'Those days are gone, I'm afraid. Still, I'm coming to see your colleagues later this month'.

'So I've heard. I'll be seeing you too. I give all the new publishers the once over. Need to test your marketing knowledge'.

'I hear Sir Henry's been testing yours'.

'Yes, and I'm not at all sure that I've passed'.

Sir Henry disengaged himself from geopolitics for a moment. 'You've passed. Not an alpha, but you're better than the others they've sent'. He turned back to the Dame of the British Empire.

'That's a relief!'

'From Henry that's the highest praise'. Lady Wiseman turned back to the publishers. 'He must want something from you'.

'We both do', interjected Sol. 'We want this paper we have just written to be published quickly. It's very important'.

'You know I can't interfere with editorial decisions. My role is purely marketing'.

'Marlene, Henry tells me you're the *burra memsahib* at

Albany. You'll be able to pull strings, surely'.

'Afraid not, at least not on that front. The Editor decides what's published and when'.

'He won't take long to decide on this one', opined Sol. 'We'd even considered sending it to *Nature*. It's that important'.

'My husband has never suffered from false modesty. Sol, shouldn't we wait until Henry has finished sorting out the latest government crisis before you speak about this damned paper. You've talked about nothing else for weeks. Fiona, I hope you've got a good lawyer'. Elvira was a dedicated student of employment law on both sides of the Atlantic.

'A lawyer?'

'If you've been fired'.

'Oh, that. Yes, thank you. One of my cousins is an employment lawyer. I think she's negotiated a pretty good redundancy package'.

'There's only one thing worse than being fired, and that's not being fired'. Marlene loved working at Albany Press and had no intention of leaving, but allowed herself occasional pangs of jealousy when she heard of the size of severance packages these days.

'Well, in this country, maybe, but not in the States'. Elvira deplored the lack of employee protection in her own country.

A lengthy discussion on employment law continued until the arrival of the sorbet, by which time Sir Henry had solved the latest government crisis, decided the Dame of the British Empire could fend for herself and turned to Marlene Pym.

'Marlene. I'm sure you can pull strings in your organization to get this paper published quickly. You're the *burra mem*'.

'So you keep telling me, but you know I can't interfere with editorial decisions'.

'Oh, but I'm not talking about that. I want you to make sure that the paper will be published as soon as possible once it's been accepted. *Acta* has got very slow indeed since I retired from the Editorship'.

Marlene knew this to be the case. Average publication times had increased from three months to six. She had complained that this was damaging the journal, but had been told that this was none of her business. In her opinion they needed a new publisher; Fiona Hamilton would be ideal.

She resorted to two excuses long beloved of publishers. 'Sir Henry, that's not my department' and 'Sir Henry, we have had very high staff turnover'.

The publishing group all laughed at this. They were only too familiar with the ritual nature of these excuses. Marlene tried to redeem herself.

'But there is one thing I can do. I can make sure that the paper is widely promoted when it is published'.

'Widely promoted? But this is science, not soap flakes. The paper will speak for itself. I don't want any vulgarity. There's been enough of that from Les Fyfe and his crew'.

'But Henry, you've got to make people aware of it', was Lady Wiseman's attempt to be constructive.

'The people who matter read *Acta* every month and will see it there'.

'The people who matter? If it really is as innovative as you claim, surely we have a duty to bring it to the attention of as many scientists as possible. Remember, *Acta* has only got 3000 subscribers'. Marlene sometimes found the *hauteur* of the grander scientists extremely hard to take.

'Henry's right. The science will speak for itself. We

don't want glossy leaflets, pop-ups, inserts, or marketing text with an exclamation mark at the end of every line', was Sol's all-too-predictable contribution.

'Don't be ridiculous, Sol. For the past month you've spoken about nothing other than this damned paper. You say it's the most important thing you've ever published, and you want to hide it under a bushel. Marketing needn't be vulgar. Why don't you listen to Marlene's ideas?' Elvira Jacobson liked to bring her husband into the real world from time to time.

Marlene smiled gratefully in Elvira's direction. 'Actually, all I was going to propose was a Press Release on plain paper, with no colour, no pop-ups, and without a single exclamation mark'.

Sir Henry and Sol were both silenced. The Dame of the British Empire turned towards the publishing group and addressed them. 'Press Release? Even the Foreign Office does Press Releases. Nothing vulgar there'.

Sir Henry needed no more persuasion. 'Well, if the F.O. does them, I suppose we could?' Sol agreed. Marlene would have her Press Release and Sir Henry even agreed to provide a suitably punchy quote to be published in it.

At the end of dinner, much to their relief, the Master's guests learned that their exit from Hall would be by a terrestrial route, which would take them to the Senior Common Room. It was there, Sir Henry promised Fiona and Tom, that they would learn the secrets of the paper that he had just submitted for publication. There, they need not worry about being overheard; no Fellow would betray a confidence given in the Senior Common Room.

There they would also enjoy the College Port, Madeira and other post-prandial delights, in relative privacy. None of the student body was permitted into the Senior Common

Room, and the college servants withdrew as soon as the assembled company were settled at the large table in the centre of the room. The final duty of the departing Steward was to light the spirit burner that powered the miniature silver steam railway engine that pulled the silver carriages containing the crystal decanters holding the drinks to be enjoyed by the Master's guests. This apparatus had been a gift of the late Maharaja of Gwalior, who had been an undergraduate at St Leonard's during the 1920s and had been responsible for setting fire to the old Senior Common Room in the aftermath of a particularly rowdy rowing club dinner. 'Squiffy' Gwalior's princely status did not prevent him being sent down, but as well as paying for the restoration of the fabric, he had commissioned this special 'Port Train' from Garrards. It had carried refreshments to thirsty College Fellows ever since.

The fact that ladies could now also enjoy the delights of the Senior Common Room and its Port Train was thanks to the incessant nagging of Lady Wiseman. This had made her Enemies among the more reactionary elements of Fellowship, but such was her force of personality that none dared betray the least discomfiture at the presence of women round their table. She, unknowingly, paid a price for her radicalism, however. One of a small coterie of Fellows known as 'The Spinsters' had a contact in Downing Street: he had made sure that her name was dropped from the list of candidates for Governor of the BBC. That would show her, if only she had been aware that her name was on the list in the first place.

The Fellowship collectively beamed as the ladies took their seats. The Port Train had built up a sufficient head of steam to set off around its silver track at a pace stately enough to permit all but the oldest Fellows to grab a decanter as it passed. As the youngest guest, Fiona Hamilton was encouraged by Sir Henry to be the first to help herself to a drink. To her surprise and delight, the train stopped when she lifted the decanter from the carriage and remained stationery until it was replaced, at which point it resumed its journey with a merry peep of its silver whistle.

It was to the pleasing accompaniment of a gently chuffing engine and the occasional silvery peep that Sir Henry and Sol revealed to Fiona and Tom the contents of their paper '*A Proposed Molecular Basis for Dementia*'.

'Sensational', was Tom's verdict. 'This opens up a whole new area of science'.

'Worrying', was Fiona's. 'If they are going to use these compounds in 'Thin-Kwik', the results could be catastrophic. We need to do something about this. Think of all these demented dogs'.

'Nothing can be done until the work is published. The science will speak for itself', was Sir Henry's considered opinion.

'We have to wait for publication and we can't risk upsetting this process'.

'Sol has a point', added Marlene, if this science is as exciting as you say, it must be published in the proper way'.

'Too right', added Tom, 'this is the most exciting work in this field in years. Correct, validated publication has to be the priority. It's the best way to answer Fyfe, Simpson and all the crap they've been publishing'.

'But what about the people who buy Thin-Kwik? They'll be thinking it will do their dogs good, when it could actually cause dementia. We've got to stop this stuff coming to the market'.

'Dr Hamilton. The best way to make sure that does not

happen is to allow the science to speak for itself'. Sir Henry was becoming lofty.

'But not everyone with a fat dog reads *Acta Moleculetica*'. Fiona was becoming sarcastic. 'It takes time for the science to percolate through to the marketplace'.

'I thought that was the purpose of the Press Release?'

'Not exactly, Sir Henry'. Marlene tried to explain to Sir Henry the purpose of a Press Release.

'Enough! You were all invited here this evening to celebrate a very significant scientific breakthrough. All I hear now is talk of Press Releases and of Thin-Kwik. If, as you say, there is a risk that this stuff will be consumed by the obese dogs of the world, the best way to prevent this happening is to make sure that the facts are made public. We scientists have done our bit. Now, it seems to me, that you business people must do yours. Ms Pym, I urge you to persuade your colleagues of the importance of this work and of the necessity to publish it as soon as possible. As Dr Hamilton has pointed out, significant harm may be done to many innocent dogs if you cannot do so. And now I shall bid you a good night and leave you in the capable hands of the Senior Fellow'.

With that, Sir Henry swept from the room.

'He didn't like you raining on his parade, Fiona'.

'I'm sorry, Sol. I didn't mean to, but if this Thin-Kwik stuff hits the market before your paper, it could do a lot of harm'.

'Henry and I are simple scientists. As he said, we've done our bit. You're smart business people. You should do yours. But don't interfere with the integrity of the scientific process. That's what caused all the trouble in the first place. Now, if you'll excuse me, I've had a long day'.

Tom broke the awkward science that 'Well, smart

business people, what are you going to do about this one?'

'They both seem very upset' observed Marlene.

'Upset! Sure they are. They're two of the most distinguished scientists of the last 50 years, they are both in their seventies, have just written one of the most important papers of their careers, invite you to dinner to celebrate, and you piss all over them. Not the behaviour expected of guests, at least out in the colonies where I come from'.

'I didn't mean to ...' Fiona was now becoming upset.

The Senior Fellow interrupted her. 'I'm sorry Dr Hamilton, no tears, it's against SCR rules'.

Marlene put a comforting arm round her. 'We'll work something out. I'll do my best to get Albany to fast track publication. It won't be easy. The journal's got a huge backlog of papers waiting to be published, and the Editor won't take kindly to interference either by me or Sir Henry. Tell me about Thin-Kwik. I'm intrigued. It sounds like the answer to a girl's prayers'.

'What?'

'Joking. You'll have to be a little less serious if you are coming to work for Albany. We're a Klomp-free zone'.

Fiona brightened up considerably at this prospect. She, Marlene and Tom retired to the suite that the Bear travel department had rented for Tom at the Randolph Hotel, where they wisely began to purge their system of vintage port, with refreshing cups of tea.

They spoke until two in the morning, covering SI, Henri de Klompenmaker, Bear, Pennine and Thin-Kwik. They agreed that they could do nothing publicly before Sir Henry's paper was published.

Fiona was rather dismissive of the proposed Press Release.

'A Press Release may sound a *petit peu* dreary, but in

273

the right hands it can be dynamite. It all depends where you send it to, and I fully intend to send this one to every media organization from the BBC down'.

'But this stuff may be on the market by the time it goes out'.

'If it takes until next June to publish the paper, perhaps, but if we fast track, we could publish as early as March next year'.

'March! That's another six months!'

'Yes, but it's going to take the Editor at least a month to have the paper reviewed before he can accept it, and that's assuming he won't want any major changes. That means it will be early December before the manuscript can go to be typeset'.

'Then, there's Christmas. Typesetters close for two weeks'.

'You see my point. It will probably be January before Sir Henry has his proofs, and you know how picky he is. He will want to discuss them with Sol. It will be the end of January before we have the final corrections'.

'Yes, yes, it looks like March is the best Albany can do. Sounds to me like your organization needs a good shake up'.

'You're telling me! That's why I think you should join us. There is something else we can do in the meantime. Marlene shifted up several gears into full marketing speed. She was now unstoppable.

'If they are to launch Thin-Kwik by March – I'm talking worst case scenario here – they'll need to have their marketing plan in place very soon. I'm talking media events, advertising campaigns, promotional mailings, even samples. For this, they'll probably be using a PR agency. I know all the major ones. And another thing: we publish some animal nutrition journals. I'll check if they've had any advertising bookings from Pennine, or if we've been invited to any new product launches. You see, there's quite a lot we can do, actually'. Fiona was encouraged. She arranged to speak to Marlene by phone two days later for an update.

'I've some good news and some bad news' were Marlene's opening words, 'which first?'

'Good news!' Fiona needed cheering up. She was missing Tom.

'Pennine have hired the Royal Yacht Britannia for a new product launch next 4th of March. Must be for Thin-Kwik'.

'The Royal Yacht? I didn't know you could'.

'It's a new thing. It seems Maggie insists that the Royal Family do more to pay their way'.

'Or she'll privatise them, I suppose'.

'Exactement'.

We were sent the brochure, but Albany are much too cheap for that sort of thing. It's available in London for a week in March, when HM's not using it. I heard that Maxwell wanted to book it for the same night as Pennine, but Pennine's PR agent sleeps with the right people, so they got it'.

'He must be furious'.

'I heard that he fired his own PR agency and threatened them with exotic forms of physical violence, as well as the usual law suits'.

'Tiffany Beale – the Pennine PR Agent – is very good. I once tried to get her for one of our product launches, but she's much too pricey for Albany'.

'What about the bad news?'

'Thin-Kwik – Pennine have applied to register it as a trademark in the UK and in the USA as a nutritional supplement *for humans'*.

'For *humans?* You're kidding. They can't possibly have done all the tests they need for regulatory approval'.

'What tests? We're talking a nutritional supplement, not a drug. Didn't you go on that SI Winter Workshop for management on the pharmaceutical industry?'

'No, I didn't think it was terribly relevant'.

'Not relevant? But the 'SI Winter Workshops' were always held in St Moritz; don't you ski?'

'Of course I ski!'

'Then anything in St Moritz in the winter is relevant'.

'I always thought that these events were just corporate junkets, and I thought I could make better use of my time'.

'So? Well, it's too late now. You'll find no 'Winter Workshops' at Albany Press. We count the bawbees here. But that's not the point. Had you gone to the workshop on the pharmaceutical industry you would have learned that one way round the endless tests that take years before you can bring a new drug to market is to call it a 'nutritional supplement' and claim no specific therapeutic effect for it'.

Fiona now, al last, understood. 'Of course. Marlene, how could I have been so stupid? That's what they're up to. We've got to stop them'.

'Easier said than done, which brings me to my other bit of news'.

'Good or bad?'

'Middling. If everything goes smoothly, we could publish Sir Henry's paper in March'.

'That's cutting things a bit fine, isn't it?'

'Yes, and as I said, it depends on everything going smoothly. The Editor will have to accept the paper quickly and without change. He'll also have to agree to putting it to the top of the queue for publication once it's ready'.

'Marlene, we've got to announce something before that'.

'And earn the undying enmity of Sir Henry Wiseman. I think not!'

'But Thin-Kwik could be very toxic. We're talking dementia!'

'Careful. That's something neither you nor I can say publicly!'

'What *can* we do, then?'

'Quite a bit, actually. To start with, I can have lunch with Tiffany Beale on some pretext; find out what's going on with this launch. I can also take the Editor of *Acta* out to lunch. I discovered that he has asked Albany Press to sponsor one of his pet projects. What with the bawbee situation here, we don't do a lot of this sort of thing, but I control the budget for it, and I might just be prepared to channel some funds in his direction'.

'A *quid pro quo*?'

'I wouldn't put it quite like that, but I can be very persuasive, you know'.

Fiona knew. 'It sounds like you have quite a lot of eating to do'.

'I know, but I'm prepared to sacrifice my figure for the good of mankind. Noble, isn't it? If only I had some Thin-Kwik, I'd be able to resist temptation'.

'Marlene!'

'Bad joke. *Desole.* Must dash now. I've got lunches to arrange. I'll keep you posted'.

Fiona was also busy. She made the most of her short time with Tom, before he left for Bear's Africa campus. Tai-Wah Chung, his trusted right-hand man, was already there, setting up the lab. After a few days in London, Tom went out to supervise the start of the research project, but was back within a month for an update on Thin-Kwik, and a romantic interlude in Venice with Fiona. During this excursion, as far as Joe Simpson and The Todd were concerned, Tom was 'in the field' gathering specimens of the exotic flora of South Africa for analysis. Weekly progress reports were faxed by Tai-Wah to Joe Simpson, who did not read them, but was much comforted by this apparent proof that Tom Carroll was nowhere near the Thin-Kwik operations in California or England.

During their trip to Venice, both Tom and Fiona had much to celebrate and plan. She had been offered a position as Publisher by Albany, and would be running their molecular sciences programme, which included the journal *Acta Moleculetica*. She would start work with them at the beginning of the new year. The fact that Albany had an office in California was an added bonus. Tom had been keeping in touch with Sir Henry on the progress of The Paper, which the Editor of *Acta* had been thrilled to receive. Sir Henry Wiseman had never been accused of taking a short-term view on anything. His horizons were both distant and broad and his mind was already turning towards the post-publication consequences of the publication of The Paper. During a discussion in Oxford he urged Tom to do the same.

'Of course, the positions of Fyfe and Simpson will be completely untenable after this is published. They will both have to resign'.

'You think so? They both have tenure and have pretty thick skins'.

'I have no doubt about it. They must both be forced to resign. Indeed I regard it as my duty to science to ensure that neither of them is ever allowed near a research laboratory again. I know how to deal with Fyfe; a word or two in the right ears will suffice. But what about Simpson?'

'I guess you could write to Dean Huntingdon. He is Dean of Research at Bear'.

'But didn't he appoint Simpson in the first place?'

'Sure, but he was misled by Fyfe. Huntingdon's a decent guy. He'll want to do the right thing. You should get in touch with the Todds. They hold the purse strings and will hate the bad publicity'.

'But I thought that the family were supporting this work'.

'Yes, Prunella Todd is completely taken in, but her husband seems more sensible. You could try him. He will not like the scandal Simpson and Fyfe will cause. They've spent millions at Bear to enhance the Todd name, not to see it dragged through the mud'.

'Prunella? That's an unusual name'.

'She's an unusual lady. She's not American. Fiona says her family own half of Scotland'.

'Aha! I knew it. She'll be Hector Maitland's girl. Bertie and I were at her wedding. I'd forgotten she married a Todd. Albany published old Hector's war memoirs. I've known him for years. We're both members of the Athenaeum. I'll have a word with him once our paper is safely published. He lives up to his name, so I think he can deal with his daughter'. What Sir Henry Wiseman did not tell Tom Carroll was that he already knew whom he would propose as the new Director of the Todd Institute. It would need a young, energetic scientist of integrity to restore its reputation.

That winter Sir Henry Wiseman gave Tom Carroll and Fiona Hamilton regular tutorials on the dark arts of academic politics. By the time the first daffodils of spring burst forth in the garden of the Master's Lodging of St Leonard's College, Oxford, his traps had been laid.

Chapter 14: The Tabloids Have their Say

The March evening was unusually balmy. The Royal Yacht Britannia, berthed in the Thames opposite the floodlit Tower of London, was herself illuminated all over. Archie Ramsbottom, Chairman and Chief Executive of Pennine Nutritional Supplements, beamed as his guests made their way up the gangway. The medley of nautical airs played by the band of Her Majesty's Royal Marines, though jaunty, made conversation difficult. Archie had told Tiffany that he wanted 'the lot' and she had delivered. Not only did he have the Royal Yacht, despite strong competition from Maxwell for that evening, but he also had the band, perfect weather and, within 15 minutes, a real live Royal Highness.

Even without the Royal Highness, the guest list was impressive. All of the major broadcasting organizations and newspapers were represented; even *The Tatler*. Tiffany was in schmooze overdrive on the aft deck and as she mingled with the media aristocracy, her tinkling laughter providing a voluble counterpoint to the band. At the bottom of the gangway, what might be termed the media peasantry were corralled behind ropes, cameras at the ready.

'Five minutes to go. They're always on time, you know'.

'Archie. You'd be on time too if they cleared the roads for you every time you set foot outside of door. Can't we go and have a drink for a few minutes until she arrives?'

'Nay, nay, we're stopping right here. See that TV camera. They'd love to get a shot of Royalty arriving and no Ramsbottoms there to greet them. We'd end up as one of those funny items they do at end of the news. I shudder to think. We're stopping right here. It's our duty. Where's Les and Vera?'

'Over there, by the bar'. Mavis pointed in their direction.

'I should have guessed. Steward, can you ask Professor Fyfe over there to step this way. He's needed'.

The composition of the Receiving Line had been a matter of much dispute among the Board Members of Pennine. Cousin Albert staked his claim as the longest serving Director, a claim that was disputed by the other Ramsbottoms. In the end, Archie decided the best thing to do was to offend them all equally by insisting that only Les Fyfe, the sole non-Ramsbottom Director, would join the Receiving Line. Les had, after all, been responsible for Thin-Kwik.

Les and Vera Fyfe downed their third glasses of champagne of the evening so far and waddled over to the top of the gangway.

'Has she arrived, then?'

'Not yet, Vera, but any minute now. They're always on time, you know'.

As if on cue, and enormous limousine pulled up at the bottom of the gangway. A lightning storm of flashbulbs erupted from the paparazzi. The door opened and the storm was extinguished in an instant amid a chorus of expletives from the media peasantry.

'F**ck! It's not Her… '. Who the hell said *she'd* be here'… 'I missed the f**king Britt Ekland restaurant opening for this'.

Tiffany Beale had, in her publicity for the event, made much of the presence of Her Royal Highness, without being very specific as to which one. To the press, there was only one Royal Highness worth photographing, and it was not this elegant, charming, but more mature woman'.

Her Royal Highness was quite oblivious to the press response, waived dutifully as she mounted the gangway, and smiled benignly as she was greeted by the Chairman of Pennine and his party.

Mavis Ramsbottom, who had been practising for days, was not ambitious in her curtsey; her size 12 dress survived the mild stresses imposed upon it by the short bob of her size 14 figure. Les Fyfe bowed and swayed slightly as he did so; a result of the champagne consumed earlier rather than any movement of the ship. Vera Fyfe, who had also been practising her curtsey, had been emboldened by the same champagne. She would show Mavis Ramsbottom how to do it properly. She sank into the lowest of curtsies, at which point gravity, that stern and unbending mistress, trapped Vera at the bottom of her sweep, Vera wobbled, her knees almost touching the deck, but was steadied by the Royal hand, which was well practised in these situations. At last, with much effort, Vera slowly began to rise, the Royal smile beaming encouragement. It was at this critical point in the manoeuvre, when a tense silence had fallen on all who watched, that a clear tearing sound began. The assembled paparazzi, with their unnerving instinct for a photo opportunity, scented blood and brought their cameras to the ready. Panicking, Vera thrust herself with all her strength back to a standing position. This final exertion proved too much for her size 16 figure to demand of her size 14 dress, which split, with a loud report and in the most revealing way, across her hips. It seemed to her that a hundred flashes went off. Tomorrow's tabloids would have their front-page photographs after all.

Her Royal Highness continued to hold onto Vera's hand while her Lady in Waiting moved forward to cover her exposed undergarments with a shawl. This manoeuvre was executed so smoothly that 'royal watchers' in the next day's press implied that Ladies in Waiting underwent special training for just such eventualities. The party moved on without further ado, the National Anthem was played and the dignity of the evening restored.

As they sat on deck afterwards reviewing the evening, Archie and Tiffany congratulated each other on its success.

Tiffany was ecstatic. 'You can't buy publicity like that! We'll be in all the tabloids tomorrow, and not just in the financial pages'.

'Poor Vera, Mavis said she just couldn't resist showing off. Too much to drink. They were both at that champagne like it were going out of style'.

'Well, she's done Thin-Kwik a favour. I must contact some of these journalists to suggest a headline to go with the photos of Vera's big behind bursting out of her frock'. Tiffany wanted to make the most of this coup. She was in PR heaven.

Archie urged restraint. 'Don't be too cruel. Pennine share price should rocket any road. All these City guys were very interested in my presentation earlier this afternoon. The shares doubled in priced before the close of play today'.

Les joined them.

'How's Vera?' Archie enquired.

'Mortified, tearful, depressed, but she'll get over it. I told her about the shares. That cheered her up'.

'I was just telling Archie that she'll be all over the tabloids tomorrow. They can't get enough of Royalty in embarrassing situations'.

'I thought Her Royal Highness handled it very well. The Lady in Waiting moved like a sprinter', said Archie.

'You know that, and I know that, but that's not how it will read in the tabloids tomorrow. We need to think of some headlines'.

'Hold on', don't you think Vera's been through enough, and you want to go encouraging the tabloids to make it worse'.

'Archie, I'm only thinking about the share price. This kind of publicity is just what we need. It's an opportunity to promote Thin-Kwik for free. The photos of Vera's backside are going to be front page news anyway. We might as well get something out of it'.

'Nay, nay, lass, I think that's going too far. Vera's been through enough'. Even Archie Ramsbottom had a chivalrous side.

Les Fyfe did not. 'I'll be the judge of that. Vera's my wife, not yours. Tiffany's given us some food for thought here. You think the right headline would really help boost the share price, do you, Tiff?'

'Without a doubt. I can't imagine anything better to draw attention to Thin-Kwik'.

'You do have a point there. Vera's backside is on the large side. What kind of headline do you have in mind?'

'That will require a bit of thought. I need a few options, you know, for the different papers. I'll come up with something. Leave it to me'.

'I don't like this at all, Les, not at all', said Archie.

'As Tiff says, the photos are going to make headlines any road, so we might as well make sure they are the right headlines. By the way, Vera must never find out about this conversation. Agreed?'

'Of course, she'll never know. Well, I must be going. I've got work to do'. Tiffany kissed Archie on both cheeks and was off.

Les and Archie waved her off. All the other guests, as well as their wives, having departed, Les and Archie made their way back to the bar, intent on having 'one for the road', while they basked a little longer in the glory of the evening.

A steward asked them where they were going.

'To the bar, laddie, where do you think?'

'I'm sorry, sir, the bar is closed. I'll have to ask you to leave the yacht'.

'Look here, young fellow, I've just hosted a big event here and I want to have a last drink before I leave'.

'I'm afraid I'll have to ask you to leave the yacht. Your time slot finished at 9 o'clock and its already ten past'.

Archie and Les's vociferous protests produced only one result. A rather burly Royal Marine who escorted them, very firmly, to the top of the gangway. No car awaited them. Tiffany had taken the last one. They made their way to Tower Bridge where they were able to hail a taxi. Neither referred to the humiliation they had just endured. Instead, they cheered each other with talk of the rising Pennine share price.

'It's more than doubled since word got out at the beginning of the week. Tiffany reckons that it could end up being worth ten times what it was a week ago'.

Les was doing two calculations in his head. One

arithmetical, the other not. The arithmetic cheered him greatly. If Tiffany was right, the profit on his stock options would be over half a million pounds; and he would be able to exercise most of them at the end of the month. He would take delivery of the new Winnebago at the beginning of April.

'We're going to be rich, Archie'.

'About time too, after all the troubles I've had with Pennine over the years. I owe you, Les, I owe you'.

Les Fyfe would usually be quick to cash in on an offer like that, but he was now distracted with his other calculation. How would he handle Vera tomorrow, when she found her bottom front page news in all the tabloids? He decided on his strategy.

'I'll make sure that she has a lie-in, so that she doesn't see the papers until we've got news of the share prices rise. That'll put her in a good mood. When she eventually does see them, I'll tell her that Tiffany tried speaking to the Editors, which she is probably doing now. Vera will assume that she tried to stop the story'.

Archie realized the folly of the words, 'I owe you', as soon as they were uttered. He braced himself for a demand. A free holiday at his villa in Spain? A company car? More share options? Nothing came and he could see that Les was deep in thought. 'He's going to ask for something big this time. He's never taken this long to get his order in before', thought Archie, as he pretended to sleep for the rest of the journey'.

When the taxi arrived at the Savoy, Archie threw a five pound note at the driver, told him to keep the change, leapt out and bade Les good night.

'What about that night cap?'

'Oh, I'm too tired. It's been a big day. You have one,

though'. Archie did not seem inclined to linger, and made straight for the elevator.

'I fully intend to', replied Les.

The next day, Les Fyfe was awoken at 7.30 by a phone call from Tiffany Beale.

'The headlines are fantastic'.

Vera scowled as she turned over. 'Who's that at this time?'

'Tiffany. It's important. Tiffany? What were you saying?'

'The headlines are fantastic. There's a picture of Vera and HRH on the front page of every tabloid, and Thin-Kwik is mentioned in every one. We were lucky that yesterday was a slow news day'.

'Oh, that's good news. Is that all?'

'Les, it's fantastic news. They even used some of the headlines I gave them. 'Her Royal Thighness' is my favourite'.

'What? Bloody hell!'

'What's wrong, Les. You don't sound very enthusiastic'.

'Oh no, it's Vera. I don't want to disturb her'.

'Sorry, Les. Anyway, you can read them for yourself. I told the concierge to deliver all the London dailies to your room'.

'To the room? Oh, God, no'.

The doorbell rang. 'Tiffany, I've got to go. There's someone at the door'.

'That'll be the papers. The staff at The Savoy are wonderfully efficient. Bye'.

Les hung up and put on his dressing gown. Vera began to get out of bed.

'Breakfast already? What's the time Les?'

'No, it can't be breakfast, it's only seven thirty. We didn't order breakfast till nine. You stay in bed. I'll see who it is'.

'You do that Les'. Vera relapsed gratefully into her pillow.

As he closed the bedroom door, Les congratulated himself on insisting that Pennine pay for a suite. Archie had been outraged by the expense, but it meant that Les could receive the pile of newspapers in the sitting room while the subject of the headlines slept next door.

On the top of the pile, 'Her Royal Thighness' screamed out above a picture of Vera's exposed undergarments. Les groaned as he sank into a chair. At that particular moment, the rise in the Pennine shares seemed a very high price to pay for Vera's inevitable wrath.

'As he went through the tabloids, he could see that Tiffany had been right. Front page news on every one'.

He moved onto the broadsheets. No photographs. No headlines. Just a discrete mention of Thin-Kwik and Pennine in the business sections. 'Exciting news from Pennine', 'Thin-Kwik a Winner', 'Buy Pennine'. These were the headlines he would share with his wife.

'Les, Les, what are you up to?'

'Oh, nothing'. He hid the tabloids under cushions. 'Just the newspapers. Very good coverage of Thin-Kwik'. He decided it was safe to let Vera see the *Daily Telegraph* and the *Financial Times*.

He took them into the bedroom.

'No *Daily Mail*? You know I like the *Daily Mail* in the morning'.

'Not this morning, you won't', Les thought. 'There's plenty news in the *Telegraph*, look.

He threw the two newspapers onto the bed, padded

back to the sitting room to check that the tabloids were well hidden, and went to the bathroom. 'As I'm up, I'm going to shave and shower'. His wife did not reply.

It was while he was shaving he heard the shriek, followed by a loud thump. He started and cut himself with his razor; he frantically stanched the flow of blood with tissue as he ran into the bedroom. His wife was out of bed and now sat on the floor, absorbed in a breakfast TV show.

'Look, look what they've done to me. I'll never be able to show my face in Rochdale again. How could they? It was an accident. I couldn't help it'.

On the screen a group of well-upholstered people on a sofa were chuckling. They had reached that part of the show where they reviewed the morning newspapers.

'My favourite is 'Her Royal Thighness. You have to hand it to the tabloids. They know how to write a headline', said the bland, portly presenter.

'I think Her Royal Highness handled it very well', added the celebrity who had been invited to review the day's newspapers. 'She's one of the low key members of the Royal Family, but she does a lot of good work'.

'She must be pretty strong. That woman she's holding onto is no lightweight'.

Vera was by now racked with sobs.

'The photographers did well to get the shot. It says here that the Lady in Waiting was quick as a flash with the shawl'.

'The best bit is that it seems the event was to launch a new slimming product'.

'Ho, ho, ho, that lady must be the 'before'.'

Vera threw herself back onto the bed and buried her head in the pillows.

'Turn it off, turn it off'.

If the Fyfes were having a difficult morning, the Ramsbottoms were having a triumphant one.

They had breakfast with Tiffany.

'Did you see the headline in The Sun?'

'I preferred the one in the 'Mirror'.

''Her Royal Thighness' was by far the best. I've already had phone calls from Lancashire. It seems it's been on breakfast TV. Poor Vera', said Mavis, with more relish than sympathy. 'I tried calling her, but Les says she doesn't want to speak to anyone today'.

'She'll come round when she sees how the share price has shot up. It's already doubled and it's not even 9.30 yet', said Archie, in comforting tones.

Tiffany, however, was not satisfied. 'We need to build on this. They're asking for more interviews and I think we should do them'.

'But we had all that yesterday. We had the *Telegraph*, the *Financial Times* and the *Independent*. What more is there to say?'

'A lot, to the tabloids. They want interviews today, with the ladies. The 'style' pages'.

'Really, with me?' purred Mavis. 'I'll need a hairdresser first'.

'And with Mrs Fyfe'.

But Mrs Fyfe would not be persuaded, and threw things at Les when he broached the subject. She wanted to get home, and fast.

In other parts of London, the morning papers were also being read with more than usual interest.

Fiona Hamilton, on her way to Albany Press's Kensington office, where she had already been working as an Editorial Director for two months, from her flat in Notting Hill, absently picked up her *Daily Mail* at the newsagents on the corner of her street.

Preoccupied with the plans for the imminent publication of Sir Henry Wiseman's paper in *Acta Moleculetica*, she did not look at the front page and exchanged only the briefest of pleasantries with the shop assistant. When she arrived at the office three messages to call Marlene Pym awaited her.

'Oh dear', she thought ' must be another hitch with this paper'.

Marlene herself picked up the phone. Most unusual, as her secretaries were usually very assiduous in screening calls.

'*Mon dieu*!' she exclaimed. *Mon dieu! Quel fiasco*'.

'Oh, no, what's gone wrong?'

'Haven't you seen the papers?'

'The papers? What do you mean?'

'Vera Fyfe. All over them. Literally!'

'Hold on, I've got the *Mail* here'. She reached for her bag.

'My God! What happened?'

'Her frock split while she was curtseying to HRH. What does the *Mail* say? I haven't seen it yet. Julia's gone out to get all the papers for me'.

Fiona retrieved her *Mail* from her bag. 'The headline reads 'House Prices Continue to Fall'. Wait a minute, here's something at the bottom of the front page 'Her Royal Kindness to the rescue'.'

'That's better than the others. The one I've got here says, in letters three inches high, 'Allo Vera; strip tease on Royal Yacht'.

Fiona read on, turning to page three of the *Mail*, which was not so nice. 'Good God, there's Vera Fyfe on page three. I see what you mean'. Fiona began to laugh. 'Our Page Three Girl Vera Reveals All'. 'How did she manage to get into that frock? No wonder it split'. Fiona suddenly stopped giggling. 'Oh no, they mention Thin-Kwik. This is disastrous. All the fatties are going to rush out and buy it'.

'You could be right there. The Pennine share price has shot up already'.

'Marlene, we've got to do something. Can't we speak to Sir Henry?'

'What good would that do? The paper is in press. We can't move any faster. You know he won't let us say a thing before it's published and we're ready to roll as soon as that happens'.

'A lot of damage could be done in the next two weeks'.

'We can't help that. Anyhow, the reaction will be all the stronger when the truth gets out. I'll be at the NASP retreat in Gstaad when the paper is published. I'll be armed with my press release. Klomp's due to speak and I'm planning to cause a sensation. I can't wait'.

At the London office of SI, Henri de Klompenmaker collected his *Financial Times*, as usual, from the receptionist. It was not a paper he read, as a matter of course, but he thought it appropriate, as a Company Director, to be seen to pick one up every morning. On this day, however, the FT was more than an accessory to Klomp's status. He read it closely.

Extremely cautious in his investment strategy, until two weeks ago, Klomp's only foray into the stock market was his 'no-risk' SI stock options, of which he had accumulated many. Then Les Fyfe, in return for getting Fiona Hamilton off his back, had tipped him to buy Pennine, a company of which he was a non-executive director. Les would let him know when to buy, as the stock, which had been falling, probably had a bit more to drop before it shot up.

'You do realize, Henri, I'm doing you a great favour here'. Les felt it was extremely important that the enormity of the favour was fully appreciated by its recipient. 'Strictly speaking, I shouldn't be doing this, but as you're a pal …'

'I appreciate the favour very much, Les. I won't forget it'.

That's what Les wanted to hear. 'Good. Now listen. I'll call you at home on the day you should buy Pennine. OK?'

'Yes, OK'.

'But to protect us both, I'll use a code word'.

'A code word? Why not just say 'Buy Pennine?'

'Because I already told you I shouldn't be doing this'.

'OK. OK. Tell me the code word'.

'Actually, it's more of a phrase. Let them eat cake'.

'Let them eat cake?'

'Yes'. Les put down the phone with a chuckle. He always enjoyed a bit of intrigue.

One evening, at the end of February, Henri de Klompenmaker's phone rang in his rented apartment in Mafeking Avenue, London. He picked it up. 'Let them eat cake'. No more had to be said, and Klomp, ever the cautious investor, devoted the rest of the evening, and much of the night, to considering whether to purchase Pennine. For the first time he actually read his copy of the *Financial Times*, saw that Pennine stock had dropped to 99 pence, threw caution to the wind and instructed his banker to purchase 2,000 shares. He now regretted this caution, as the price climbed steadily. By lunchtime, the price of Pennine was six times what he had paid.

In the parallel universe that is St Leonard's College, Oxford, the Wisemans, who did not watch breakfast television and did not allow tabloid newspapers to cross the threshold of the Master's Lodging, were listening to the morning news on BBC radio as they read the *Daily Telegraph* and the *Guardian* in the sunlit morning room. The name 'Pennine', which rang a bell with Sir Henry, featured prominently in the business pages and was even mentioned on the radio news. He furrowed his brow, but could not make the connection and passed on to an intriguing article about an elderly Englishman rumoured to be living as a member of a native tribe in the Amazon jungle. News of Vera Fyfe's discomfiture on the Royal Yacht would not penetrate the Master's Lodging for some days.

In Montecito, California, Prunella Todd ate breakfast on her terrace and scanned the pile of faxes that Desmond had brought to her. She passed each page to her husband as she finished it.

'Franklin, this is most exciting. Look at these newspaper headlines that Joe Simpson has faxed. The launch of Thin-Kwik has been a huge success. The Royal Yacht and a Royal Highness! I knew we should have gone'.

'But you said that the Britannia was nothing special'.

'It's not, at least compared with Forbes' yacht, but they should have told me that HRH would be there. Mummy will be furious'.

'Prune, who's that woman collapsing in front of the Princess?' Franklin pointed at the grainy faxed photograph.

'Oh, I don't know. It's probably some stunt, you know, to show the embarrassing situations being fat can get you into. Here's another picture of her. Wait a minute'. What's this headline? 'Allo Vera? Franklin, it's her, that English professor's wife. Oh dear, how dreadful, in front of Royalty too'. Prunella began to laugh uproariously.

'Looks like HRH came out of it rather well … and her Lady in Waiting seems to have saved the day … Lady Jean Boyle'.

'What, cousin Jeannie? I didn't know she was on duty. How wonderful. She'll be thrilled. Let me look at that one again … Of course, she always was nimble on her feet. Mitch, come and see this'.

Mitch, who was hovering in the background waiting for an opportunity to review the schedule for that day with her mistress, was at the table in an instant.

'Mitch, look at this. Cousin Jeannie's a hero'.

As her personal maid looked through the faxes with mounting fascination, Prunella bade Desmond fetch her husband's secretary.

'What do you want Joyce for?'

'I've got faxes to send and phone calls to make. Mummy can't know about this, or she'd have called. And I must fax the girls. They're already green with envy about Compound X, or should I say Thin-Kwik. They should know about the Royal Yacht and the Princess. I tell you, Franklin, the Todd Institute is the best hundred million dollars you've ever spent'.

Joyce arrived, notebook in hand.

'Joyce, good. I'm going to need you this morning'. Franklin frowned 'Just for a couple of hours. First, can you get Lady Maitland on the phone for me? Then, I'll want these faxes sent on to some people. Mitch, haven't you finished reading them yet?'

Mitch looked up. 'I'm reading this stuff about the share price going up because of Thin-Kwik'.

'That'll be Cheviot, I suppose. I told you to buy the shares, Franklin'.

'Pennine', countered Mitch.

'Honey. I told you, we couldn't, it would be insider dealing. We'd have broken the law'.

Mitch, who heard everything and forgot nothing,

blushed. She remembered something urgent she had to attend to and handed the pile of faxes to the secretary. 'If you're going to be busy this morning, I'll get on'.

'Yes, Mitch. I'll be busy till lunchtime. You'll have to take Ko Ko to the vet yourself. Let's go Joyce'.

Mitch, despite her considerable bulk, could move very quickly when necessity demanded. She made her way across the garden to the cottage that was her private, Pekinese-free demesne. Her parrot Clara, a refugee from Prunella's considerable flock, enjoyed classical music and liked to listen to Public Radio while Mitch was busy during the day. She resented this unaccustomed intrusion on her morning routine.

An irritated 'What do you want', greeted Mitch.

'Never you mind'. Mitch called her stockbroker.

'Sell, sell!' cried the parrot, who sensed when Mitch was involved in one of her regular excursions into high finance. Mr Todd had given her frequent, profitable tips over the years.

Mitch: 'How's Pennine doing?'

Stockbroker: 'Reached a new high today. Over £6 a share in London'.

'Should I sell?'

'Not yet. The word is that this one has a way to go. The institutions are beginning to buy big time'.

'Yes, but I heard …' Mitch stopped herself. 'Overheard' would have been more accurate, and Fiona's suspicions remained with her.

'Look, this is my grandson's college fund. I'll hold a bit longer, but sell as soon as it hits £10 a share'.

'At this rate, it could reach that tomorrow. Are you sure?'

'Listen. A ten-fold profit is good enough for me. Sell

when it hits ten'. Spoken like a pro.

Her stockbroker knew better than argue with Mitch when she said 'buy' or 'sell'. Her instinct, inspired as it was by Franklin Todd's tips, had proved very reliable over the years. Sell at £10 it would be.

The Pennine stock reached the magic £10 on the following Monday. Mitch's stockbroker acted on her instructions and sold, turning Mitch's original $10,000 investment into just over $100,000 dollars.

'Another tidy profit for Mrs Mitchell', he mused . 'How does she do it?' But Mitch seemed to be the only one selling, and the stock had reached new heights by the end of the week.

Chez Fyfe the tension was mounting. Vera had not left the house since she and Les returned to Rochdale from London. Their intended low-key departure from the Savoy by the service entrance had been sabotaged by the paparazzi, and a second day of unpleasant headlines ensued. She dared not show her face in Rochdale. As far as Vera Fyfe was concerned, there were journalists lurking under every shrub. To restore her spirits, Les promised that they would take to the road in the new Winnebago as soon as it was delivered, and encouraged her to phone their broker twice per day to check on the Pennine stock. She passed on the bulletins to Les at his lab.'Mid-morning price: £11.20'; 'It hit £12 this afternoon'.

Chapter 15: The Slippery Slope

By the time Henri de Klompenmaker stepped onto his flight for Geneva at the end of the week, he was in agonies of indecision. 'Should he sell, or should he hold?' The Pennine price was now hovering at just over £12. If he sold his entire holding he would have to pay capital gains tax. But what if the price fell? On the other hand, some newspapers were predicting that it could reach twenty. After another sleepless night, he was no closer to making up his mind. Despite having attended management, finance and strategy courses of every description, he could not match the decisiveness in such matters of an elderly Scottish domestic servant.

So tired and distracted was he that he even failed to recoil from Marlene Pym when she greeted him with a cheerful '*Bonjour*, Klomp' as she passed up the aisle to her seat.

'Not again!' he groaned as the rising chatter from a few rows behind provided evidence that Marlene was accompanied by her usual band of cronies. Klomp congratulated himself on insisting he have a limo from Geneva to Gstaad. His secretary, who had lived in Geneva and claimed to be an expert on all things Swiss had tried to persuade him to go by train. No way. He had never forgotten the embarrassment of waiting for a taxi in Bermuda while Marlene and friends swept off in their limos. No more public transport for Henri de Klompenmaker.

The flight was a short one, and apart from the background noise of their constant chatter, Marlene and her group did not impinge upon him. To take his mind off Pennine, he looked through his presentation. It would build nicely on his triumph of the previous year. *Transactions in Moleculetics* had progressed well since then, and the results he would present would be even more spectacular. That would silence Marlene Pym. There would be no awkward questions this time.

Their arrival in Geneva was met with the usual Swiss efficiency. The baggage was already on the carousel by the time they were through passport control. As there was no sign of Marlene, Klomp assumed she must have gone ahead already. This was wildly optimistic of him.

'No skis, Klomp?' She was standing behind him with a porter in tow.

'What?'

'Skis. If you've checked skis, you need to pick them up over there. They don't come through on the carousel'.

'Skis? No'. He thought he detected her exchange knowing glances with her assembled cronies.

'*Pas de skis*? You're aware that this is Switzerland? And that they have the best spring snow in years'.

'I'm here to work'.

'Aren't we all, darling? Aren't we all? Joining us in the first class carriage on the train to Gstaad, are you?'

He knew that she was baiting him. His put down was ready.

'Train? Ha! Ha! You must be joking. Train? Can't you

do better than that Marlene? I've got a limo to Gstaad. Albany hitting hard times, are they? Ha! Ha!'

This response was beyond Marlene's wildest dreams. She surveyed her attentive audience before replying.

'Klomp. Nobody goes to Gstaad by car. Nobody. Train, yes, Plane, yes. Helicopter, yes. Car, no. I'll let you in on a secret, Henri. Switzerland has mountains'.

'I know that very well'.

'The point about mountains is that they have snow in winter, you know. Snow that blocks the mountain passes for months on end'. Marlene was drawing out the agony, much to the appreciation of her onlookers. 'They had very heavy falls of snow last week. Lots of roads still closed'.

'So what?'

'So what? It means you'll have to drive half way round Switzerland to get to Gstaad. It will take you five hours at least'.

'Very interesting. Just as well I have a limo. Have you any more useless information for me?'

'Only this. We are all off to enjoy one of the great scenic rail journeys of the world, during which we shall enjoy first class comfort, food and drinks, and which will get us to Gstaad in under three hours. Bon voyage, Klomp'.

They departed, in the highest of spirits, followed by three porters laden with assorted bags and winter sports equipment.

Five and a half hours later, the limousine bearing a rather weary Henri de Klompenmaker pulled up beneath the portico of the Belvedere Hotel in Gstaad.

'Welcome to the Belvedere, Sir' was the porter's cheerful greeting as he opened the limousine door. 'The snow's wonderful, as I'm sure you've seen on the way up'. His colleague was already unloading bags from the trunk.

As they walked into the lobby, the porter tried to make small talk. 'Did your skis not make it, Sir? They are often delayed, I'm afraid. You can hire skis at the hotel until yours arrive'.

'I don't ski. I am here to work. I'm attending a conference'. Klomp was not in the mood for polite small talk. The porter continued to try his best.

'Ah, yes. Your colleagues arrived earlier. Some of them made straight for the pistes. Wonderful snow'.

'So you keep telling me'.

The porter did not receive a tip and Klomp replied in monosyllables to the conversational gambits of the desk clerk during check in. He was particularly irritated to be told that the reason for the absence of guests in the hotel was that everyone was out on the slopes enjoying the wonderful snow.

His room had magnificent views across the rooftops of Gstaad to the meadows, pine forests and ragged peaks beyond. Everything was clad in a blanket of snow, and the whole picture was turning a soft pink in the early evening sun. On the pistes the last skiers were making their final runs and on the hotel terraces chattering groups were gathering for an après ski drink. On the terrace of the Belvedere Hotel, a particularly chatty group was gathered around Marlene Pym, her silver/grey ski suit now glinting in the setting sun.

'How did you manage to get this place at such a good rate, Marlene?' was the question on many lips.

Marlene sipped her gin and told her tale.

'Well. The thing is, the few weeks between February and Easter is actually the low season here, and as the hotel has to stay open, they may as well fill it up. My aunt and uncle have been coming here for years and know the Kotmans

– that's the owners – very well. I called Maria Kotman and did a deal. We're paying less than 50% of the full rate'.

'We'll make up for that in the bar', quipped the President of the Academy.

Marlene went on to explain that they don't normally do conferences at the Belvedere, for fear of disturbing their regular, pampered, wealthy guests. But she had been able to convince Maria Kotman that the annual retreat of the Newtonian Academy of Scholarly Publishing was not like a normal conference. Only a few lectures each day, no commercial exhibition and no badges. The vacationing Crowned Heads of Europe would not be disturbed.

'We can eat all our meals in the dining room, so long as we sit at tables of no more than eight people and spread ourselves about', Marlene elaborated.

'I trust that we can order *a la carte*, as usual?' an anxious listener asked.

'Naturellement. Set menus at NASP? Unthinkable! By the way, the wine list here is fantastic'.

To nods of appreciation all round, Marlene described the many other delights of the Belvedere. She herself had already made an appointment at the spa for a caviar facial treatment.

Buffy Cavendish, unusually for him, had a question about one of the lectures. 'Has Lola Santiago arrived yet? I'm looking forward to her talk. Didn't understand a word last year, but she makes you think, don't she?' Burbles of masculine appreciation greeted this remark.

'Yes, and I know exactly what she makes you chaps think about. She arrived last night and I am told has been heli-skiing with her personal ski instructor all day'.

'She'll be joining us for dinner, then, eh?'

'Yes, Buffy, she will. But this could be the last year.

We've used up all the reserves to pay her fee. We'll have to get someone cheaper next time'.

'Nonsense. Good talk don't come cheap. We'll have to start charging a supplement'. There was a general and vociferous agreement with this remark and the talk moved on to the afternoon's skiing.

Henri de Klompenmaker was as unaware of this discussion as he was of the spectacular view from his room. He was on the phone to London to find out how the Pennine stock was doing. It was doing well, he was told, and still hovering at just over £12. He agonized. His banker suggested he might sell half now, which would lock in a substantial gain with minimal tax consequences. Klomp was not sure. He would sleep on it and decide tomorrow. He needed to focus on his presentation now. As he was tired after his lengthy journey, he went downstairs only briefly, to pick up his conference information pack. Waiting in the lobby for the elevator, he saw a number of familiar figures, clearly enjoying themselves, drinking on the terrace. He made up his mind to order dinner in his room.

He also decided to breakfast there the next morning, to give himself another opportunity to go over his twenty-five slides and make sure that he was word perfect. Marlene Pym would not catch him out this time. A lot had happened since his presentation last year. Other publishers had taken up his management techniques, although it had to be said that their application was rather unsophisticated. This year the results for *Transactions in Moleculetics*, his favoured example, were even more spectacular. More citations, more papers and more profits. Yes, SI had much to thank Henri de Klompenmaker for and he expected this to be reflected in his next promotion.

That night, with so much on his mind, he slept very

badly. During the opening, 2-hour morning session of the Newtonian Academy's retreat, he kept dozing off. He even fought hard to suppress his yawns during Lola Santiago's plenary lecture. Her ill-advised remarks at the SI strategy meeting had shown her to be a guru with feet of clay. Even the tight leather pants she wore this morning could not hold his attention. By the time the final speaker, an egghead from CERN in Geneva, rose to address them on '*The Internet and the Future of Publishing*', Klomp was fast asleep and was roused only by the rapturous applause at the end of this talk. He decided he would rest in his room that afternoon, making sure that he was fresh for his own presentation in the evening. Consequently he missed the excited chatter about the potential of the Internet, over lunch.

By the time Henri de Klompenmaker strode into the Baroque Room of the Belvedere Hotel, he was completely refreshed. His afternoon nap had worked its magic, and he had also had another opportunity for a final review of '*A Quantitative Approach to Journal Management: Part Two*' before he spoke. To his chagrin, he had not been able to speak to his banker. He had decided to sell half his Pennine stock, but the line was permanently busy. For the first day in two weeks he had no information on how Pennine was doing. This was a great irritation, and Klomp privately cursed his absence from London at this time.

The room was full, just as it had been in Bermuda. This time, however, there was no Lionel Grove to marshal the audience. They had come of their own accord. Marlene Pym had promised some first class entertainment and they had hastened down from the pistes to be in place for the five o'clock start of the session. The first two presentations generated only one or two questions.

Klomp was introduced, stood up and began to speak.

In his customary, mechanical way he systematically went though each slide. As usual, he was virtually unaware of the audience, and was entirely absorbed in his word-perfect script. He finished slide twenty-five exactly on time. The Chairman asked if there were any questions. Only one hand went up and all eyes turned towards Marlene Pym.

'Mr de Klompenmaker' – she was always very correct on these occasions – the results you have presented for the *Transactions in Moleculetics* are very interesting'.

Klomp: 'Thank you'.

Marlene: 'The continuing increase in citations and impact seem impressive'.

Klomp: 'They are impressive'.

Marlene: 'The growth is based on the high number of citations to a relatively small number of papers, are they not?'

Klomp: 'Yes, but that's not unusual'.

Marlene': A small number of papers from a single institution in California'.

Klomp: 'Correct. Excellent papers from a first class institution'.

Marlene: 'All short communications without details of how the experiments were done'.

Klomp: 'Yes, but short communications are quite normal practice in this field of science'.

Marlene: 'But it's also normal practice to follow them up with full papers that contain the details of the experiments carried out. This allows others to reproduce the results, does it not?'

Klomp: 'Yes'. Klomp knew there were no scientists present, so he thought he would get away with his next response. 'But these results are from a very sophisticated laboratory. The experiments need expensive equipment, and this is not widely available. Only a few institutions can do this kind of advanced research'.

Marlene: 'That's a very pompous remark, if I may say so'.

Klomp: 'It's also true. Most places couldn't afford the equipment'. He had no idea whether this was, in fact, true, but he also knew that nobody in the Baroque Room was in a position to contradict him. He would tough this out.

Klomp: 'The Todd Institute is big and very expensively equipped. It's the best'. He searched for a metaphor to describe the famed lavishness of the Todd Institute's facilities. 'It's like the, er, the Versailles of moleculetics. Most labs don't have the funds to do this sort of very advanced research. Pity for them'.

Marlene leapt at the opportunity to take Klomp's metaphor further: 'In other words, '*Qu'ils mangent de la brioche*?'

'What?' This response was not the 'what?' of incomprehension; Klomp spoke fluent French. Rather, it was the 'what?' of surprise, but he could not explain this to his interlocutrix.

'Let them eat cake'. Marlene was relishing the amusement of her audience and did not notice the colour drain from Klomp's cheeks.

Her modest little tease had quite taken the wind out of his sails. She must know something.

Marlene took advantage of the pause to press on with her questions: 'Would it be reasonable to describe the reaction mechanisms proposed in these short communications as controversial?'

Klomp: 'Novel is more accurate, but I can't comment on that. I'm not a scientist'.

Marlene: 'No, but Sir Henry Wiseman is a very

distinguished scientist, would you agree'.

Klomp had heard Les Fyfe rail at length against Sir Henry Wiseman, but knew little of him: 'I believe so, but what's that got to do with it?'

This answer provoked the first open laughter from the audience. Everyone, apart from Klomp, knew that Sir Henry Wiseman, Nobel Laureate, was one of the most famous of living scholars. They had all seen his landmark TV series *Knowledge* in the 1970s. All apart from Henri de Klompenmaker, that is.

Marlene: 'Let me tell you what that's got to do with it. Today, in *Acta Moleculetica*, which is published by Albany Press, a paper has been published by Sir Henry Wiseman and Sol Jacobson that disproves entirely the reaction mechanisms proposed in these short communications you have been publishing in *Transactions* these last few years'. Marlene dramatically raised her arm. 'And I have here a press release announcing publication of that paper which, by the way, contains full details of the experiments done'.

Klomp tried to keep his composure: 'There's often controversy over new theories and reactions. It's the normal scientific process. Lots of debate and discussion'.

Marlene: 'It is bad enough to publish flawed, unsubstantiated science in your journal, but to hold it up as a model for managing the scholarly publishing process is an insult to everyone in this room. It is disgraceful. You are bringing publishing into disrepute. Fie, Klomp, fie'. With this final expression of disgust, Marlene sat down, well satisfied with her peroration, but then she immediately stood up again, addressing the room in general. 'By the way, I have more copies of the press release here. It was released in London this morning'.

Klomp attempted to regain the audience's attention,

but nobody was listening. The session ended in chaos as Marlene Pym was mobbed. Copies of her press release fluttered into the air as Klomp slipped out and hastened to his room. He had other, more pressing matters on his mind. He had to sell these Pennine shares, and quickly. Marlene Pym knew something. 'Let them eat cake, indeed'. If he was found holding shares he could be charged with insider trading. Luckily, he had not bought many, so if he sold them now, he might get away with it. He checked his watch. Seven o'clock! Too late! The London market would be closed. His banker would have gone home, but he could still call Miel Flick, and Les Fyfe. Waiting for the elevator further, more alarming thoughts crowded about him.

'If that press release went out this morning ... the Pennine shares ... the journal ... my career'.

He tried his banker first; he might be working late. No answer. He tried Miel; she was usually in the office at this time. No answer. He tried Les Fyfe's home number. Answering machine. He lay down on his bed and began to comprehend the scale of the unfolding disaster.

Marlene was having much more success on the phone. After things calmed down in the Baroque Room, she resisted the temptation to go with the flow to the bar. She wanted an update from Fiona Hamilton in London, excused herself with a promise of more news later, and made for her suite.

'Fiona? Marlene here'.

'Marlene! All hell's broken loose'.

'But earlier on you said there had been little reaction to the press release'.

'It took a while to build. At first we got just a few phone calls from the scientific press, you know, *New Scientist* and the like. They recognized Sir Henry's name. Then we started getting calls from the science editors of the dailies. We put the *Telegraph* onto Sir Henry. Remember how he said he would only speak to the qualities; no tabloids. That was how it started. It seems they asked him about the connection with Thin-Kwik. He replied 'No Comment'.

'Fatal'.

'Exactly! The next thing, we're being called by one of their City correspondents'.

'What time was that?'

'About one o'clock. I tried calling, but the hotel couldn't find you. Didn't you get the message?'

Marlene blushed as she recalled her lengthy, sun-kissed lunch on the piste.

'Oh, I was in a meeting. You know what it's like. What did you say to the City correspondent?'

'All his questions were about Thin-Kwik. 'Is is true that it may cause dementia? Has it been fully tested? What did we know about Pennine? Did I know why the share price had started to tumble that morning?' I said that I couldn't possibly comment'.

'Wonderful. I bet that got him going'.

'It certainly did. Especially when I mentioned that the person he should really talk to is Professor Leslie Fyfe, who, I understood, was both Editor of *Transactions* and a Director of Pennine. I think he smelt a rat'.

'*Formidable*! And what next?'

'Well, I told him that I really didn't have any further information, and he should probably speak to Professor Fyfe himself, who I was sure would want to clarify things. I did mention that I happened to have Professor Fyfe's phone numbers, if he wanted them'.

'And?'

'He couldn't write them down fast enough'.

'Great! Anything else?'

'One thing puzzles me. Why did the Pennine share price fall so quickly after the press release went out? It didn't even mention Pennine or Thin-Kwik'.

'Tiffany Beale! *Absolutment!*'

'Tiffany Beale? The PR? How come?'

'She was on the fax list for the Press Release, wasn't she?'

'Of course'.

'Remember, she's good and will know all about the Bear/Pennine connection. When I had lunch with her she was very cagy about it all'.

'Yes, you told me all she would say is 'Let them eat cake'. Marlene had forgotten that, but now remembered Henri de Klompenmaker's reaction when she had used the phrase earlier that evening. This got better by the minute, 'Marlene, are you still there?'

'Yes, about Tiffany. She'll have seen the words 'disproves the reaction mechanism proposed by Simpson *et al.* of Bear State University, published in a series of papers in *Transactions*'; alarm bells will have started ringing. I bet she faxed it onto Pennine within ten nanoseconds, where the Directors will have taken appropriate action'.

'By selling their shares?'

'Exactly. Pennine will soon be worthless. Think of all the people who have bought Thin-Kwik in the last fortnight. The law suits alone will bankrupt them. Now tell me, what's the latest price?'

'That's my last bit of news. Trading in Pennine has been suspended, after the price fell to £1'.

'A 90% drop in one day? Not bad!'

'One thing galls me, Marlene. If what you say is true, I mean about the Directors selling shares this morning, Les Fyfe has probably made a lot of money'.

'*Cher* Fiona, if that's the case, he'll probably go to jail along with the other Directors for insider trading. Look on the bright side'.

'That is cheering news. He should really go to jail for scientific fraud, but financial fraud will do. How are things at your end?'

'*Pauvre* Klomp', purred Marlene in *faux* sympathy. '*Humiliation totale*. Everyone is talking about it. I practically got trampled to death in the stampede for the press release'.

'So he's been found out'.

'And how! He must have sneaked out in the *grand debacle* at the end. There's going to be mega-fallout from this'. Marlene was eager to participate personally in the fallout. 'Look, I must get back down stairs; I've got people waiting for meetings'.

'I bet you do. Speak to you tomorrow. *Bon Chance!*'

'I beg your pardon?' Marlene had come to regard the use of French as a linguistic garnish as her own personal trademark.

'*Bien sur*. Speak to you tomorrow, Marlene'.

That same evening, *chez* Fyfe, as Marlene might have put it, the telephone rang and rang, but was not answered.

'Take the damned thing off the hook', called Vera from the darkened bedroom in which she had taken refuge. 'I've got a splitting headache'. She did not enjoy the fame that the evening on the Royal Yacht had brought to her doorstep. The tabloids were lurking outside again.

Her husband ignored both his wife and the telephone as he nursed a whisky in their living room with the curtains drawn, only a flickering TV for company. This evening he wanted to converse only with himself. He had accumulated quite a stock of grievances since that telephone call earlier in the day from the *Daily Telegraph* City correspondent.

'Bloody Archie Ramsbottom. I bet he made a pretty penny when he sold all those shares this morning. Not Les, though, not Les. Oh no, Les's Director's options can't be acted on until the first of April. Who's the bloody April Fool, then? Les, good old Les'. He had tried calling Archie Ramsbottom several times in the course of the day, but Archie, acting on Tiffany Beale's advice, declined to speak to him and was by now on a flight to Malaga. He and Mavis would lie low at their Spanish villa while the dust settled. The five million pounds they had made from the sale of only 20% of their Pennine shares this morning would pay the bills for a while. In fact, it was more than their entire stake in Pennine had been worth before Thin-Kwik. So what if the rest were now worthless? They had been unpleasantly surprised to bump into several other Ramsbottoms at Manchester Airport that evening, all bound overseas.

Les continued to review his shit list. 'And that old bastard Wiseman. He's had it in for me for years. Thinks he's God's gift to science. One word from Wiseman of Oxford and the world turns upside down. Well, if they believe that old charlatan, good luck to them!' That evening, the old charlatan and his wife were dining alone in the Master's Lodging. In the interests of domestic harmony, Sir Henry had not mentioned his conversation earlier that day with the City correspondent of the *Daily Telegraph*. It was a newspaper of which Lady Wiseman thoroughly disapproved, a distaste, she had once assured him, shared by Marshall Stalin. Sir Henry was called from the table during dinner to take an international phone call. On his return, he informed his wife that it had been Sol Jacobson who sought to confirm that their paper had actually been published in *Acta Moleculetica* that day. They both anticipated with relish the stir that it would cause in the academic world in the coming months.

'Prunella bloody Todd. More money than sense. Her and her kind should be shot. If it hadn't been for her damn fool son getting himself gored by a buffalo, we could have all lived happily ever after. Todd Institute? Bad joke!' It was afternoon in California. Prunella Todd was hosting her regular, weekly bridge party at her Montecito mansion, Elysium. The bridge was taken seriously, but the exchange of gossip even more so. Mitch, almost $100,000 richer since last week's game, busied herself in the next room, the door slightly ajar. Today the ladies chatted about one of their favourite topics, weight loss. Three of them expressed their astonishment at the way Shelley Borman had shed weight over the last few months. They also expressed their irritation at her clearly diminishing powers of concentration. She had become a liability as a bridge partner. Mitch, absorbed in the same bridge game, was ambushed by Franklin Todd, who apologised for disturbing her but wanted to let her know that his advice against buying Pennine shares had been justified. They were now crashing and his broker had told him a rumour was circulating that the new Pennine wonder supplement caused dementia. Mitch's eyes widened steadily as he told his tale, and as he left, wishing her 'Good listening', she crept up to the door and examined Shelley Borman very closely indeed.

A late entry in the Fyfe Villains Hall of Fame was the Manchester Winnebago dealer. 'What do they mean, we can't cancel the order. Bloody ridiculous. It's a glorified caravan. Bespoke my arse, any one would think it was a damned Rolls Royce. Produced to my specification, bollocks. It's on its way from the States already? That's just to wind me up. I've signed a contract, have I? Forty thousand quid!

Forty thousand quid!' The Winnebago was indeed on its way across the Atlantic and barring a repeat of the Titanic disaster, Les would be obliged to part with £40,000 within a matter of weeks.

'As for de Klompenmaker, he can get stuffed. He'll want to make me the fall guy. Keep his nose clean. I'll be the sacrificial lamb, just like Fiona Hamilton. How does that guy get away with it? He knows nothing. Ignorant berk'. But, in Gstaad, as a restless evening turned into a sleepless night, Henri de Klompenmaker had a growing sense that this time he was not going to get away with anything. The sacrifice of Les Fyfe was a foregone conclusion, but that might not be enough. 'What if this affected the SI share price?' He sat bolt upright in bed. In his obsession with Pennine, he had forgotten all about SI. Further promotion would be out of the question now. He'd be lucky to hold onto his present job. At least his status and pension would be safe. His contract of employment was under European law, not English, and he was virtually unfirable. But he could well follow other failures to SI's own private Siberia, the East European branch of the Natural Resources Division. Klomp paced up and down, his mind turning, turning, turning. He decided that he would check out tomorrow morning, after calling his banker and Miel. He couldn't stand another day of this retreat; Marlene Pym would be insufferable. So absorbed was he in his thoughts he did not notice that outside the snow was falling heavily.

Chapter 16: Marlene Has Her *Mot*

Marlene Pym did not feel her usual chirpy self when she awoke next morning. Her triumph had been too well celebrated and she now had a hangover.

'7.30. Good, this morning's session does not start until 9.30'.

She called the hotel spa. 'Rapid detox and rehydration treatment? That's exactly what I need … No, I don't mind sharing, I suspected there would be heavy demand for your services today'.

In a less exalted part of the Belvedere Hotel, an irritated Henri de Klompenmaker shouted down the phone. 'What do you mean I can't leave today? Gstaad cut off by snow? Biggest blizzard in 50 years? No I haven't looked out of my window'. He looked. He hung up.

'Shit. Stuck in this hole!' It was still only seven o'clock in England. Too early for him to call London, but apparently not too early for London to call him. His phone rang.

'Henri de Klompenmaker? Chairman's office. Mr Archer would like to speak with you most urgently'.

'De Klompenmaker! What the hell's going on? I read in

the papers this morning that we, a leading scholarly publisher, have no idea what we're doing, that we'll publish anything, and that we've been taken in by a bunch of charlatans. Do you know what this will do to our share price when the markets open today? Just when we've had this fiasco in the Natural Resources Division. And I thought I could count on the Publishing Division. What the hell are you doing in Switzerland anyway? Another jaunt, I suppose. Get the first flight back. I want this sorted out'. The Chairman hung up before Klomp could get a word in.

Henri de Klompenmaker needed a drink. Breaking the habit of a lifetime, he opened the minibar. A beer would do nicely. And then another. It was still only 8.30 in England. He was still thirsty. He opened a miniature of gin while he waited. He switched on the TV and flicked through the channels. He stopped at CNN Business News. '… and in London, yesterday's spectacular fall in Pennine shares is the main talking point. Trading in the shares was suspended yesterday afternoon when they fell to £1. It's hard to believe that only three days ago, these shares were worth over £12. Police in England are searching for Mr Archibald Ramsbottom, Chairman of Pennine, who is rumoured to have left the country. Other Directors, who are believed to have dumped large quantities of shares just before the crash, are also being sought … and it doesn't stop there; shares in Standard International, the leading international conglomerate, have also taken a tumble when trading opened today in London, following confirmation from Brazil that the elderly Englishman found living with the natives in the Amazon jungle this week is, in fact, Sir Edmund Jackson, the distinguished scientist who disappeared several years ago. In a news conference scheduled for later today, Sir Edmund is expected to claim that while he has been living in hiding he has been collecting evidence on illegal logging operations by the Standard International Corporation. Even more sensationally, he is alleging that SI tried to have him murdered, and claims this is why he went into hiding. CNN will be broadcasting the news conference live from Brazil, so stay tuned.....'

Klomp could stand it no longer. He switched off the TV and opened a second miniature of gin. He gazed mindlessly out of his window as the snow continued to fall. At last, nine o'clock, English time, arrived and he called his office. He told his secretary to have Miel phone him when she got in.

'But Miel's been moved. She was called yesterday by the Chairman's office. Moved to Corporate Strategy'.

He called his banker, who confirmed that trading in Pennine was suspended and expected there would be a further plunge when trading in the shares started again. In fact, they would be worthless.

Having finished the gin in the minibar, Klomp moved onto the vodka and continued drinking.

There was a larger than usual group in the Belvedere spa that morning. Marlene, quite revived by her 'rapid detox and rehydration treatment', was holding court.

Prescient as always, the hotel management had instructed their London agent to scan the first editions of the English dailies and fax copies of every paper mentioning Pennine, Thin-Kwik or SI. Twenty bound copies of these faxes awaited Marlene and the others when they arrived at the spa.

'Isn't Swiss service wonderful?' opined Marlene to general agreement. 'They anticipate everything. The concierge has also arranged for us to receive a bulletin on the Pennine and SI share situation every hour throughout the day. I'm sure nobody will mind the interruptions'. Nobody did.

'The pistes are closed, so we shall have to entertain ourselves indoors instead. By the way, has anybody seen Lola this morning? She'll love this'.

By the time the time came for the freshly detoxed, rehydrated conference participants to make their way to the Baroque Room for the start of proceedings, they felt fully briefed on the situation.

The exciting climax of Henri de Klompenmaker's professional suicide on the previous day could have left that morning's session feeling a bit flat, in the manner of the final scene of Lucia de Lammermoor, the heroine herself having made her spectacular, blood-stained exit at the end of the one before. But the session turned out to be remarkably lively. The detoxed and rehydrated participants were intellectually turbocharged and the announcements about Pennine and SI at the end of each presentation brought an additional thrill. Marlene had also arranged for the Albany share price to be given to her privately. It was rising steadily.

As the morning progressed, Henri de Klompenmaker worked his way through the contents of the minibar and sunk deeper into gloom as he contemplated the ruins of his publishing career. At around eleven, he was aroused by the tap on the door that signalled the arrival of Carmen the maid to make up his room. He told her to come in and pretended to watch the television as she busied herself vacuuming, dusting and polishing. He replied only in monosyllables to her polite comments on the weather until she said that the trains from Gstaad were running again.

'Running again? Too late; he couldn't be bothered leaving today. Nigel Archer would just have to wait until tomorrow. If there's any truth in what this Jackson guy is saying, he's probably finished anyway. Klomp was by now enjoying this unaccustomed decadence. He was even thinking of ordering a bottle of champagne to see him through the afternoon. Thus minded, his eye fell on Carmen's ample, rounded behind as she bent over the bed to change the pillow cases. It was all too much. Klomp's moment of madness would be the talk of the Belvedere for the remainder of the retreat. Nobody had thought him capable. But they were unaware of the years of sexual frustration that lurked inside. In his own mind, Klomp was still the acned youth of twenty years ago. He was normally too timid to take the initiative with women, but today, fuelled by an excess of alcohol, Carmen's swaying buttocks proved irresistible. He grabbed them. Outraged, she turned and punched him on the face, her uniform tearing in his hand as she did so. Seeing him fall to the floor, she screamed and ran for the housekeeper.

Within five minutes, both housekeeper and maid were in Maria Kotman's office. Within fifteen, Marlene Pym had been called from her meeting to discuss a most delicate matter regarding the behaviour of a NASP delegate with Madame Kotman.

'Ms Pym, this is very embarrassing', began Maria Kotman. 'It appears that one of our maids may have been assaulted by one of your delegates'.

'No, surely not!' Marlene responded emphatically, but without conviction. She knew only too well the sorry history of previous NASP Retreats.

'I'm afraid so. Her dress is torn and her hand is bruised. She is hysterical'.

'Poor thing, but are you quite sure?'

'Carmen is from Spain. She doesn't speak any other language very fluently and with the state she's in, we can't understand what she's saying'.

'Perhaps I can help you there. We have a Spanish

speaker at our conference. Dr Lola Santiago. She could talk to Carmen'.

'That would be very helpful. Shall we send someone down to fetch her?'

'Oh, I haven't seen her in the conference room today. Perhaps she's out skiing. Somebody saw her with a ski instructor earlier'.

'But all the pistes are closed. The heavy snow last night. She must be somewhere in the hotel. I'll ask the bell boys to have a look'. Maria Kotman called the front desk. She then went briefly into the next room to check on Carmen, who was being comforted by the housekeeper.

On her return, Marlene wanted to know more. 'Now, about this delegate. Have you any idea who it might be?'

'We're not sure, but Carmen's trolley was outside a Mr De Klompenmaker's room'.

Marlene instantly became even more sympathetic to the maid's plight. 'De Klompenmaker!'

'You know him?'

'Oh, yes, vaguely. Who'd have thought …'.

Marlene's reveries were interrupted by Maria Kotman's phone. It was the front desk.

'It seems that Dr Santiago is in her suite, and the 'Do Not Disturb' sign is up. This is most unfortunate'.

'Never mind that. This is an emergency. I'll go to her suite myself. Show me the way'.

Marlene's banging on the door provoked much stamping of feet and hostile comments from within, but once appraised of Carmen's plight Lola's mood changed and she appeared in her dressing gown. Before closing the door behind her she called into the room 'Hey, Sven. Don't go. I'll be back in no time'.

Maria Kotman was not at all surprised that Dr Lola

Santiago appeared in her office in a bathrobe. This was a resort hotel, after all. She was taken to Carmen immediately.

'Leave us alone' were Lola's only words in English. She put her arm around Carmen and began to speak softly, in Spanish. The others tiptoed from the room After fifteen minutes Lola Santiago emerged, closed the door softly, stamped her feet and launched stream of Spanish expletives.

Marlene calmed her down, and Carmen's tale was told.

Lola's advice was unambiguous. 'You must call police immediately. This creep assaulted her'.

Maria Kotman did not like this idea at all. At the Belvedere they took care of such unfortunate incidents more discretely; by doing so had earned the gratitude and custom of Crowned Heads of Europe. 'These things are so difficult to prove. Swiss justice takes for ever. I think that instead Mr De Klompenmaker should be asked to leave the hotel immediately'.

'No, police must be called immediately. If you no call, I call'. Lola stamped her feet very emphatically.

Marlene, mindful of the Newtonian Academy's reputation, agreed with Maria. 'Hold on, Lola. Even if he goes to jail, what punishment is that. Swiss jails are like five star hotels. If, on the other hand, he is thrown out of the Belvedere in full view of the entire publishing world ...'

'Total humiliation! I like it. You English are so subtle, and so cruel!'

'Thank you, Lola'.

Marlene assured Maria that the Newtonian Academy would quite understand that no other course of action was possible. Before she and Lola left the room, Marlene asked at precisely what time Maria planned to have Mr de Klompenmaker escorted from the hotel.

Klomp was in his bathroom when he heard the knock on the door, observing with some alarm the growing swelling and discolouration around his left eye, where the maid had punched him.

'About time', he thought. 'They've come to apologise'.

He opened the door, to be confronted by Maria Kotman, the housekeeper and two porters. Maria walked straight past him into the room, followed by her entourage.

'Mr de Klompenmaker', she began, 'I have just been told by one of our maids that you assaulted her in this room'.

'What? She must be fantasising'.

'Fantasising, about you? I think not! Carmen has worked for us for twenty seasons. She has looked after the crowned heads of Europe. Her character is unimpeachable'.

'I didn't … she's made it all up. Look what she did to me!'

'Made it all up! I have seen her torn uniform myself'.

'Torn uniform?'

'Yes, torn. I cannot allow my staff to be treated in this way. I am afraid I shall have to ask you to leave the hotel'.

'I think you are over-reacting. She's only a maid'.

'A maid whom I value highly. Unless you leave this hotel within the hour, I shall call the police and charges will be pressed. Do I make myself clear?'

'Charges? An hour? But I'll have to dress and pack'.

'These gentlemen will take care of your packing. A car will be waiting for you at the front door at noon. We shall have your bill ready for you to sign. Dominique'. She was now addressing the housekeeper. 'Would you be so kind as to check Mr De Klompenmaker's minibar. It looks like there may be a few drinks to be added to his bill this morning'.

'But where will I go, what will I do?'

'Frankly, that is up to you, but I suggest that you see a doctor about that eye. I'll arrange for the car to take you to the local clinic before you leave Gstaad. Meanwhile, why don't you take a shower and make yourself decent while your bags are being packed'.

Maria Kotman then left him alone with three very hostile domestics. He decided to take her advice and locked himself in the bathroom to wash and shave. His left eye was by now completely closed.

In Rochdale, the Fyfes were having an equally miserable morning. Fed up with the growing press encampment outside their front door, they decided to leave town to sit out the storm with Vera's sister in Bradford. They were not on speaking terms. Having exhausted their considerable stock of mutual recriminations by the end of the previous day, they had nothing left to say. In complete darkness they had dressed and in the early morning gloom they had slunk out of their back garden gate, hoping to get away unnoticed in Vera's little runabout. Banks of cameras flashed the moment the gate closed behind them.

At noon, a little procession left room 304 of the Belvedere Hotel, descended in the elevator and crossed the lobby on its way to the waiting car. In the lobby stood the Fellows of the Newtonian Academy, who watched in silence as the procession paused and Maria Kotman stepped forward. 'Your bill, Mr de Klompenmaker. I trust you will find it in order. We have already charged it to your credit card'.

Klomp accepted the bill without a word. Dr Lola Santiago spat at his feet as he passed her . Looking neither left nor right, he followed the porters towards the front door, where stood Marlene Pym.

'*Qu'ils mangent de la brioche*, Klomp?'

He blanched. Let them eat cake? Again? She must know something. Exactly how much she did know would become clear in the coming weeks and months, as various investigations got under way.

About the Author

Tom Angus is intimately acquainted with the academic and publishing worlds on both sides of the Atlantic. In the course of his career he has greatly enjoyed his relationship with the world of scholarship and has been hugely entertained by the excesses of corporate life. He has been both a perpetrator and a victim of the publishing process. His publications include many articles in journals, as well as chapters in reference books, but *Let Them Eat Cake* is his first novel. Tom Angus was born and lives in Scotland.

Printed in the United States
57810LVS00001B/6

9 781420 810486